Nancy Freitag

The role of galectin-1 during preeclampsia

Nancy Freitag

The role of galectin-1 during preeclampsia

The pro-angiogenic function of galectin-1 in healthy and pathological pregnancies

Südwestdeutscher Verlag für Hochschulschriften

Impressum / Imprint
Bibliografische Information der Deutschen Nationalbibliothek: Die Deutsche Nationalbibliothek verzeichnet diese Publikation in der Deutschen Nationalbibliografie; detaillierte bibliografische Daten sind im Internet über http://dnb.d-nb.de abrufbar.
Alle in diesem Buch genannten Marken und Produktnamen unterliegen warenzeichen-, marken- oder patentrechtlichem Schutz bzw. sind Warenzeichen oder eingetragene Warenzeichen der jeweiligen Inhaber. Die Wiedergabe von Marken, Produktnamen, Gebrauchsnamen, Handelsnamen, Warenbezeichnungen u.s.w. in diesem Werk berechtigt auch ohne besondere Kennzeichnung nicht zu der Annahme, dass solche Namen im Sinne der Warenzeichen- und Markenschutzgesetzgebung als frei zu betrachten wären und daher von jedermann benutzt werden dürften.

Bibliographic information published by the Deutsche Nationalbibliothek: The Deutsche Nationalbibliothek lists this publication in the Deutsche Nationalbibliografie; detailed bibliographic data are available in the Internet at http://dnb.d-nb.de.
Any brand names and product names mentioned in this book are subject to trademark, brand or patent protection and are trademarks or registered trademarks of their respective holders. The use of brand names, product names, common names, trade names, product descriptions etc. even without a particular marking in this work is in no way to be construed to mean that such names may be regarded as unrestricted in respect of trademark and brand protection legislation and could thus be used by anyone.

Coverbild / Cover image: www.ingimage.com

Verlag / Publisher:
Südwestdeutscher Verlag für Hochschulschriften
ist ein Imprint der / is a trademark of
OmniScriptum GmbH & Co. KG
Heinrich-Böcking-Str. 6-8, 66121 Saarbrücken, Deutschland / Germany
Email: info@svh-verlag.de

Herstellung: siehe letzte Seite /
Printed at: see last page
ISBN: 978-3-8381-5056-7

Zugl. / Approved by: Berlin, FU, Diss., 2014

Copyright © 2015 OmniScriptum GmbH & Co. KG
Alle Rechte vorbehalten. / All rights reserved. Saarbrücken 2015

Table of Contents

1. Introduction 1
   1.1 Reproduction in mice and humans 1
   1.2 Galectin-1 8
   1.3 Galectin-1–mediated adaptions of the immune system to pregnancy 11
   1.4 Galectin-1 and angiogenesis 12
   1.5 Pregnancy complications and galectin-1 16
   1.6 Aim and main results of the thesis 20

2. Methods 22
   2.1 Laboratory equipment and consumables 22
   2.2 Reagents 25
   2.3 Buffers and solutions 30
   2.4 Antibodies 31
   2.5 Animal models 32
   2.6 Experimental sampling 33
   2.7 Human samples 34
   2.8 Immunohistochemistry and immunofluorescence 36
   2.9 Protein isolation and Western blot analysis 39
   2.10 Angiogenesis array 40
   2.11 RNA isolation, RT-PCR and quantitative real time PCR 41
   2.12 Enzyme-linked immunosorbent assay (ELISA) 44
   2.13 Angiotensin II receptor type 1 autoantibodies 45
   2.14 Kidney function assessment 46
   2.15 Blood pressure measurements 47
   2.16 *In vitro* assays 48
   2.17 Microarray 48
   2.18 Statistical analyses 50

3. Results 51

3.1 Gal-1 acts as a pro-angiogenic factor during early gestation in mice 51

3.2 Inhibition of gal-1–mediated angiogenesis provokes PE-like symptoms 60

3.3 Gal-1 inhibition impairs human trophoblast functions *in vitro* 67

3.4 *Lgals1* deficient mice spontaneously develop PE-like symptoms 70

3.5 Placental gal-1 is dysregulated during PE in humans 74

4. Discussion 78

5. Summary 86

6. References 89

7. Publication list 98

8. Appendix i

8.1 Abbreviations i

8.2 List of figures iii

8.3 List of tables v

8.4 Description of Theiler stages vi

8.5 Selection: Dysregulated genes in human trophoblast microarray vii

1. Introduction

1.1 Reproduction in mice and humans

Reproduction is fundamental to sustaining life and although the development of a placenta requires a high metabolic investment, eutherian mammals have been especially successful during evolution. The advantages of a placenta lie in the extended time *in utero;* enabling a long maturation period before the foetus is born and a high activity state that the mother can maintain during pregnancy. While mice produce many offspring that reach sexual maturity after a few weeks, humans developed a strategy of producing few offspring that can be well protected and nourished.

One of the main differences between mouse and human reproduction is the length and physiology of the cycle. The oestrous cycle of mice lasts approximately four to five days [1]. During the follicular, pro-oestrous and oestrous phases, several follicles in the ovaries grow and uterus vascularization increases. Under the influence of rising oestrogen levels, the epithelial and stromal cells in the endometrium (inner cell lining of the uterus) proliferate. A progesterone surge occurs right before ovulation at the beginning of the oestrous phase, the female is receptive for fertilization during the next hours, and the morning after fertilisation is denoted as gestation day (gd) 0.5. In the following luteal phases (met- and di-oestrous), corpora lutea form in the ovary to secrete progesterone, which promotes the differentiation of stromal cells in the endometrium. Moreover, leukocytes start to infiltrate the stroma in the met-oestrous phase, while the vascularization and number of epithelial cells in the uterus decrease. If no conception occurs, the corpora lutea degenerate in the di-oestrous phase and a new cycle starts. However, if the eggs are fertilised, they migrate through the oviduct to the uterus, meanwhile developing to blastocysts consisting of an undifferentiated inner cell mass (ICM) and a surrounding trophectoderm (Figure 1) [2].

Introduction

Under the influence of progesterone from the corpora lutea and oestrogen from ovarian follicles, the uterus is receptive for blastocyst implantation from gd 3.5 to 5, a period called implantation window [3] (Figure 2). Embryo attachment to the endometrium triggers the decidualization of the surrounding endometrial cells [4]. During this process, the stromal cells transform into decidual cells and the extracellular matrix is remodelled under the influence of oestrogen and progesterone [5]. The differentiated endometrial tissue is now called decidua and develops into anti-mesometrial and mesometrial decidua. In the anti-mesometrial decidua, the embryo has implanted and the decidualization process is initiated; the mesometrial decidua is characterised by angiogenic processes and is the designated site of placentation (Figure 2). The decidua supports embryo growth and maintains pregnancy until the placenta undertakes the task of supplying the embryo with nutrients and ensuring gas and waste exchange.

Eutherian mammals are characterised by a chorioallantoic placenta that is formed by foetal and maternal membranes [6]. The chorion is the outermost foetal layer and consists of epithelial cells that are derived from the trophectoderm. The underlying extraembryonic allantois comprises endothelial cells that line the capillaries and connective tissue. Primates and rodents develop a discoid-shaped, haemochorial placenta during gestation [7]. Since the maternal layers (endothelial cells, connective tissue and epithelial cells in the endometrium) are transformed during placentation, the maternal blood comes into direct contact with the foetal chorion. However, morphogenesis and some endocrine functions of the placenta differ between mice and humans.

In mice, the definitive structure of the placenta is established during mid-gestation [2,8]. Trophoblast stem cells differentiate into proliferating trophoblasts of the extraembryonic ectoderm and ectoplacental cone, which later gives rise to the either the spongiotrophoblast of the placenta or (by endoreduplication) to primary trophoblast giant cells (GC) (Figure 1). The first

Introduction

stage consists of a choriovittelline placenta with a parietal yolk sac as the first placental structure from gd 6 to 8 (Figure 2). It forms between the primary GC and the basement membrane, which is made from migrated parietal endoderm cells. Nutrients are absorbed from the maternal blood via small capillaries that later form sinuses. Subsequently, migrated mesoderm cells form another placental structure, the visceral yolk sac (Figure 2). These cells differentiate to the first vascular cells and thereby establish the vascular zone on gd 7.5.

In the next days, the allantoic mesoderm contacts the chorion of the ectoplacental cone. Thereby, the definitive structure of the chorioallantoic placenta is formed on gd 10.5 and remains until the end of gestation (gd 19 to 20). The chorioallantoic placenta consists of a labyrinthine and junctional zone, separated from the maternal decidua basalis by a GC layer (Figure 2). The labyrinth is established by invading trophoblasts that fuse to form syncytiotrophoblasts (STB) and enable an efficient exchange between maternal and foetal blood vessels. The junctional zone contains spongiotrophoblasts and glycogen trophoblast cells (Table 1). In the last years much progress has been made in the identification of transcription factors that determine the cell fate of the stem and differentiated trophoblast cells [9].

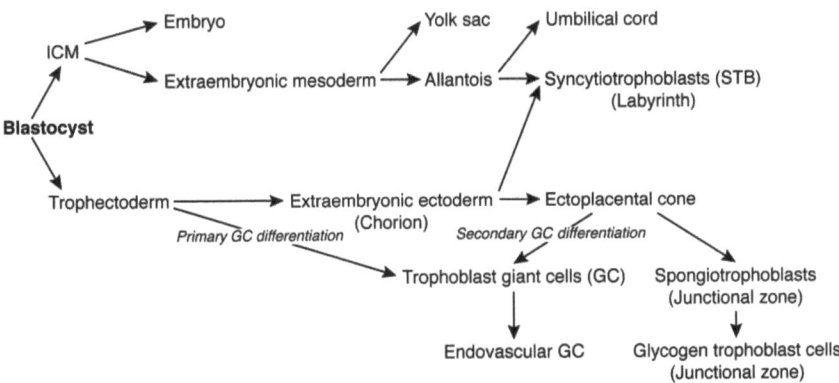

Introduction

Figure 1: Differentiation of trophoblast cells from the inner cell mass (ICM) and trophectoderm in mice
The definitive structure of the placenta is established at gd 10.5 and comprises five differentiated trophoblast cell types: syncytiotrophoblasts (STB), trophoblast giant cells (GC), endovascular GC, spongiotrophoblasts, and glycogen trophoblasts [2,9,10]. The STB differentiate from trophoblasts in the allantois and chorion. The first GC derive from the trophectoderm in a primary differentiation during early gestation. The main pool of GC in the placenta stems from a secondary differentiation of trophoblasts in the ectoplacental cone.

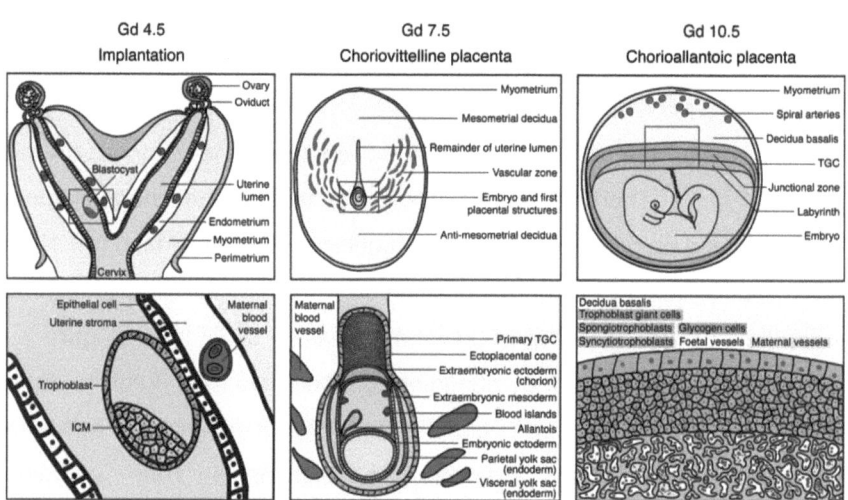

Figure 2: Placentation in mice
After fertilisation, the zygotes migrate from the oviducts to the uterus and develop into blastocysts that consist of an inner cell mass (ICM) surrounded by trophoblasts. The duplex mouse uterus has two horns and the uterine wall is formed by endo-, myo-, and perimetrium. Around gestation day (gd) 4.5, the blastocysts attach to the epithelial cells and implant in the stroma of the endometrium. While the epithelial cells undergo apoptosis, the stromal cells decidualize. At gd 7.5, a mesometrial and anti-mesometrial decidua can be distinguished. A parietal yolk sac forms as a first transport organ at the mesometrial site and is referred to as choriovittelline placenta. Subsequently, a visceral yolk sac forms and gives rise to the vascular zone. The definite placental structure establishes at gd 10.5 and is called the chorioallantoic placenta. Maternal spiral arteries in the decidua basalis are remodelled to ensure the blood supply of the foetus. Trophoblast giant cells (TGC) separate the decidua basalis from the underlying placental structures. Spongiotrophoblasts and glycogen trophoblast cells form the junctional zone. The labyrinth is composed of syncytiotrophoblasts that mediate the exchange between maternal and foetal blood vessels. The functions of the differentiated trophoblasts are summarised in Table 1.

Table 1: Trophoblast cell functions in the mouse placenta

Trophoblast cell type	Function
Trophoblast giant cell (GC)	Uterus invasion during implantation [9] Promotion of embryonic growth: angiogenesis of maternal blood vessels [11] and endocrine functions (production of growth factors, placental hormones) [9]

Introduction

	Immune evasion of foetal antigens [2]
Endovascular GC	Invasion of maternal decidua and remodelling of spiral arteries [12]
Spongiotrophoblast (Junctional zone)	Structural function [9]
	Expression of anti-angiogenic factors (sFlt-1 [A], Prp [B]) to prevent growth of maternal blood vessels into foetal part of placenta [13-15]
	Endocrine functions [16]
Glycogen trophoblast cell (Junctional zone)	Interstitial invasion of decidua during later gestation, but no invasion of spiral arteries [12]
	Function unknown, but glycogen-rich granules indicate a nutritional function for the growing embryo [17]
Syncytiotrophoblast (STB) (Labyrinth)	Barrier between maternal and foetal blood [9]
	Transport of nutrients within placenta [9]

[A] sFlt-1: soluble fms-like tyrosine kinase-1, reduces the bioavailability of VEGF;
[B] Prp: Proliferin-related protein, antagonizes proliferin

The human menstrual cycle lasts about 28 days. In the menstrual and proliferative phases, ovarian follicles mature until the oestrogen level peaks and causes a surge of luteinizing and follicle-stimulating hormone that induces the ovulation of usually one egg around day 14. In the subsequent secretory phase, the corpus luteum produces oestrogen and progesterone. If the egg is fertilised, the levels of these two steroid hormones remain high and do not decrease in the late secretory phase. In the oestrogen-primed uterus, progesterone induces the differentiation of endometrial stromal cells into decidual cells [18]. The cellular and biochemical properties of decidual cells promote the potential implantation and later also the growth of the embryo [5].

The blastocyst implants 8 to 10 days after ovulation and produces human chorionic gonadotropin (hCG) to preserve the corpus luteum and thus progesterone production until the placenta has developed [19]. The trophoblasts of the blastocyst penetrate the epithelium and basement membrane and contact the decidualizing stroma of the endometrium [20,21]. At this stage, the primitive trophoblasts of the trophectoderm already differentiated into pre-villous syncytiotrophoblasts (STB) that locally degrade the extracellular matrix and form the first lacunae, which later expand to the intervillous space (Figure 3) [20].

Introduction

In the third week after conception, the chorion consists of a monolayer of mononucleated villous CTB stem cells that differentiated from the primitive CTB and forms the first villi that increase the chorionic surface and float in the intervillous space. The villi are covered with a continuous layer of multinuclear STB that form by fusion from the underlying CTB stem cell layer and later ensure nutritional exchange with the maternal blood. Foetal blood vessels appear in the mesenchymal core of the chorionic villi. In some villi, the CTB stem cells break through the STB layer and make contact with the maternal decidua, thereby forming trophoblast cell columns and anchoring the placenta to the decidua. These extravillous CTB differentiate into interstitial and endovascular CTB and are characterised by an invasive behaviour (Figure 3, Figure 4).

In the first trimester, the interstitial extravillous CTB start to migrate into the decidua and some cells differentiate into multinucleated trophoblast GC with endocrine and nutritional functions. In the second trimester, the interstitial extravillous CTB invade until the myometrium underlying the decidua. Additionally, endovascular extravillous CTB replace endothelial and smooth muscle cells of the maternal spiral arteries, resulting in low resistant, dilated blood vessels that guarantee a sufficient blood supply to the placenta and embryo. Although the spiral arteries are remodelled during the first half of pregnancy, the trophoblasts continue proliferating and the placenta grows throughout pregnancy. The placentation process is finished by the end of the first trimester as defined by the definitive structure of the placenta and flushing of the intervillous space with maternal blood [8].

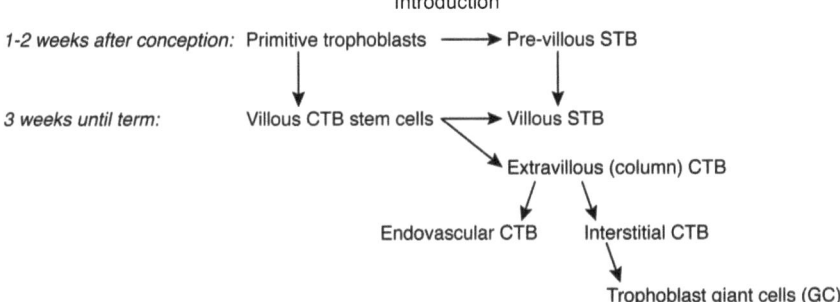

Figure 3: Differentiation of human trophoblasts
Human trophoblasts differentiate from the primitive trophoblast cells of the trophectoderm in the first two weeks after conception. Pre-villous syncytiotrophoblasts (STB) are important for the very early placentation processes including degradation of the maternal extracellular matrix and formation of lacunae in the uterus. After 3 weeks post-conception, the villous stem cell cytotrophoblast (CTB) pool has formed and continuously proliferates until term. These cells give rise to the villous STB and extravillous interstitial and endovascular CTB that invade the maternal decidua and spiral arteries. James et al. suggested that there are two different villous stem cell CTB pools from which the STB and extravillous CTB derive [22]. Some of the extravillous interstitial CTB differentiate into trophoblast giant cells (GC) within the maternal decidua [20].

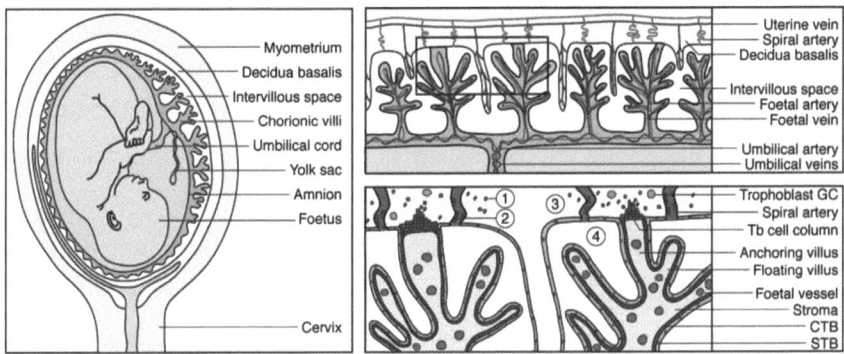

Figure 4: The human placenta
Until the placentation process is finished in the first trimester of human pregnancy, the yolk sac supplies the embryo. In the definite structure of the placenta, maternal spiral arteries deliver blood to the intervillous space, in which the chorionic villi are bathed. The chorionic villi increase the surface area for the uptake of nutrients and oxygen by multinuclear syncytiotrophoblasts (STB). STB fuse from the underlying cytotrophoblast (CTB) stem cell layer and cover the anchoring and floating villi. Foetal vessels in the mesenchymal core (stroma) of the villi transport the nutrients to the embryo via the umbilical cord. In anchoring villi, extravillous CTB break through the STB layer and form trophoblast (Tb) cell columns. Interstitial extravillous CTB (1) invade the decidua basalis (3) with some differentiating into trophoblast giant cells (GC). Endovascular extravillous CTB (2) invade the spiral arteries and remodel them into dilated blood vessels for an adequate blood supply to the intervillous space (4).

1.2 Galectin-1

1.2.1 General aspects

Galectin-1 (gal-1) belongs to the evolutionarily conserved family of β-galactoside binding lectins that are characterized by their carbohydrate recognition domains (CRDs). It was the first protein discovered and is considered as a prototypic member in the family of galectins [23]. A human gal-1 monomer has a molecular mass of 14 kDa and folds into a β-sandwich that consists of two anti-parallel β-sheets of five and six strands [24]. With the two CRDs located at opposite ends, the homodimer exhibits a typical "jelly-roll" folding; in solution the shape is maintained by non-covalent, hydrophobic interactions. The gal-1 monomer and homodimer exist in a reversible equilibrium and are associated with different biological activities [25,26].

Intracellular gal-1 is involved in protein-protein interactions and modulates processes such as cell growth, differentiation, transformation, migration, survival, and pre-mRNA splicing [26,27]. The lectin activity of gal-1 also plays an important extracellular role by binding to integrins, fibronectins, laminins, and osteopontin [26]; suggesting a role in cell-cell and cell-matrix interactions [28-30]. Because gal-1 lacks a secretion signal sequence, it is secreted into the extracellular space via non-classical direct translocation across the plasma membrane [31,32,33].

The *LGALS1* (lectin, galactoside-binding, soluble 1) gene locus consists of 4 exons that encode a protein of 135 amino acids. *LGALS1* maps to human chromosome 22q13.1 and the mouse *Lgals1* gene is located on chromosome 15. Detailed investigations of the *LGALS1* gene showed a gain of 10 *cis* elements in the evolutionary tree of eutherians [34]. Some of these regulatory elements in the *LGALS1* promoter can be induced by oestrogen: a half-oestrogen responsive element (ERE), nuclear transcription factor (NF)-Y and activator protein (AP)-2 [34-36]. Interestingly, ovarian steroids like oestrogen and progesterone increase endometrial gal-1 levels [37]. The blocking of oestrogen and progesterone nuclear receptors and progesterone deficient mice also display decreased gal-1 levels showing the importance of maternal gal-1 during gestation [37,38].

1.2.2 Maternal expression of gal-1 during reproduction

In all mice – sexually immature, ovariectomized, or at any stage of the oestrous cycle – gal-1 is expressed in all uterine tissues except the glandular and luminal epithelium [39]. The investigation of gal-1 localisation and expression patterns in the maternal cell subsets helps to understand gal-1 function during the different stages of pregnancy. Before and during implantation, the cytoplasm and some nuclei in the myometrium and endometrial stroma are positively stained [39]. The decidualized stroma cells still display a strong staining in their cytoplasm and nucleus. The blood sinuses in the vascular zone also express gal-1. During the days after decidualization, the gal-1 levels in the anti-mesometrial decidua decrease. In contrast, the decidual and uterine natural killer (uNK) cells in the mesometrial decidua, where angiogenic processes take place, still express high gal-1 levels. After the establishment of the decidua basalis, all nuclei of the natural killer (NK) cells are stained, whereas the staining of cytoplasm is variable. In advanced pregnancy, gal-1 expression remains unchanged, but glycogen trophoblast cells, which invade the decidua basalis, display a strong gal-1 expression [12,39].

Like steroid hormones, gal-1 expression fluctuates during the menstrual cycle. In the human endometrium, gal-1 is expressed during the proliferative and early and mid-secretory phases and is up-regulated in the late secretory phase [40]. Gal-1 levels are further increased when the endometrial stromal cells become decidualized in the first trimester supporting the idea of an essential role for gal-1 during pregnancy. The glandular epithelial cells and leukocytes in the late secretory endometrium and decidua exhibit only low gal-1 RNA and protein levels. In contrast, the stromal cells have high levels of gal-1 and thus account for the increase in gal-1 during early pregnancy [40-42].

At term, the uterine glands, leukocytes and maternal blood vessels in the decidua are negative, while the stroma is positively stained for gal-1 [42].

1.2.3 Early embryonic and placental gal-1 expression

In mice, gal-1 is expressed in the trophectoderm of the blastocyst, while expression is absent in the ICM [43]. Due to this particular expression pattern, and the fact that gal-1 mediates cell-cell interactions implicates gal-1 in the implantation process (embryo attachment and the invasion of the uterine epithelium by trophectoderm) [26,43]. At the beginning of mid-pregnancy while the embryo proper is still negative for gal-1, expression is seen in the ectoplacental cone, suggesting an involvement in trophoblast differentiation [39]. After completion of the placentation process, the spongiotrophoblasts and STB express gal-1 in their nuclei and cytoplasms [39,44]. The vascular spaces in the labyrinth are also positively stained. This indicates that gal-1 could support exchange processes within the placenta by mediating cell-cell interactions. The GC display a weak cytoplasmic gal-1 staining with negative nuclei and gal-1 is detected in the connective tissue of the myometrium. Twenty-four hours after parturition, the gal-1 distribution resembles the virgin uterus with gal-1 expression in smooth muscle cells, the myometrium and endometrium, but not in the luminal epithelium of the uterus [39].

Gal-1 is expressed during early human development at the cleavage and blastocyst stage in the ICM and trophectoderm, which, as assumed in mice, suggests an involvement in the implantation process [26,41,43]. Furthermore, gal-1 expression in the trophectoderm argues for its function as a key protein in the differentiation of trophoblasts. Indeed, the villous CTB in the first trimester of human pregnancy express gal-1 and syncytium formation is stimulated by gal-1 *in vitro* [41,45]. Interestingly, the differentiation into STB is accompanied by a loss of gal-1 expression [41,42,46]. In contrast, the extravillous CTB are also differentiated from villous CTB, but display an increased gal-1 expression [20,41]. James *et al.* [22] suggested that STB and extravillous CTB are derived from two distinct stem cell pools of villous CTB, which might be supported by the differential gal-1 expression in these trophoblast populations and their different functional properties: syncytium formation versus invasion of maternal tissue [41]. The interstitial extravillous CTB display the highest gal-1 expression in the trophoblast cell lineage, localised both perinuclear and extracellular [42,46-48]. Moreover, the villous mesenchymal cells and the walls of the foetal blood vessels in the chorionic villi display an expression of gal-1 throughout pregnancy [42].

1.3 Galectin-1–mediated adaptions of the immune system to pregnancy

Paradoxically, though the foetus expresses paternal antigens, it is not rejected by the maternal immune system [49,50]. While the mother is still protected from infections, immunological processes in the decidua are tightly regulated. Indeed, many pregnancy complications are associated with disturbances in immune tolerance mechanisms [51]. The tolerogenic and anti-inflammatory effects are well studied in the adaption of the maternal immune system to pregnancy. First of all, gal-1 is expressed in the decidua and maintains the balance between T helper type 1 (T_H1) (TNF-α, IFN-γ, IL-12) and T_H2 cytokines (IL-4, IL-5, IL-10) [52-55]. In a mouse model of stress-triggered abortions, foetal loss was increased due to increased pro-inflammatory, abortogenic T_H1 cytokines [52,53]. Decidual gal-1 levels were decreased in these mice, and supplementation with exogenous gal-1 prevented foetal loss and rescued pregnancy success. It is hypothesized that gal-1 restores the T_H1-T_H2 cytokine balance via the stimulation of tolerogenic CD11c$^+$ dendritic cells (DC); tolerogenic DC subsequently promote interleukin (IL)-10 production by regulatory T cells (T_{reg}) and T_H2 cytokines that thereby suppress alloreactive responses of the maternal immune system [52,56-59]. In accordance with this, *Lgals1* deficient mice have an exacerbated T_H1

response, their DC have a greater immunogenic potential and they are more susceptible to stress-induced abortions [60-62].

In humans, tolerogenic signals are derived from gal-1 expressing uNK cells that constitute the main leukocyte cell subset infiltrating the maternal decidua [63,64]. Gal-1 from uNK cells induces apoptosis of decidual activated T cells in order to protect the foetus from being attacked by the mother's immune system [65]. Moreover, uNK cells have been reported to shape the immunological functions of DC in mice, which subsequently induces the expansion of T_{reg} and NK cells and increase IL-10 concentrations [66,67]. A functional DC-NK cell crosstalk is necessary for the maintenance of pregnancy, however, the direct involvement of gal-1 in the DC-NK cell crosstalk remains to be elucidated.

Besides these maternal adaptions, foetal trophoblast cells also have an indispensable contribution to the maintenance of pregnancy [51]. Trophoblast cells express complement regulatory proteins to avoid maternal antibody-mediated cell lysis, and B7 family proteins protect foetal cells against activated maternal leukocytes [68,69]. The expression of human leukocyte antigens (HLA) by trophoblasts plays a pivotal role, since parents differ at their HLA loci and the mother produces antibodies against paternal HLA inherited by the foetus. Nevertheless, the foetus is not rejected, because HLA-G targets most immune cell subsets and binds killer inhibitory receptors (KIRs) on NK cells [51,70]. Gal-1 regulates the HLA-G expression in the placenta and thus contributes to the immune evasion strategies of trophoblasts against the maternal immune system [41].

1.4 Galectin-1 and angiogenesis

1.4.1 Angiogenesis during pregnancy

Angiogenesis is the process by which new blood vessels arise from pre-existing capillaries and its temporal coordination is crucial for a successful pregnancy. Angiogenesis is involved in follicular and corpus luteum development during the reproductive cycle and enables the implantation of the blastocyst into a highly vascularized receptive uterus [71]. During decidualization, the endometrial vasculature extends, and new blood vessels form, to supply the embryo with oxygen and nutrients before the placentation process is finished. Normal placental development is highly dependent on proper angiogenic signals to fulfil the rising demands of the growing foetus. The importance of angiogenesis was demonstrated in female pregnant mice that were treated with the anti-angiogenic drug AGM-1470 [72]. A single dose of AGM-1470 prevents implantation or placentation depending on the gestation day of administration.

Trophoblast-derived signals are especially important for pregnancy-associated vascular changes. Trophoblast giant cells (GC) in the mouse placenta are in direct contact with the maternal decidua and produce angiogenic and vasoactive factors such as pregnancy-specific glycoprotein (PSG) 22, proliferin (PLF), vascular endothelial growth factor (VEGF), and adrenomedullin to promote the maternal blood flow [14,73-77]. In humans, it has been shown that extravillous CTB promote decidual cells to express many vascular factors *in vitro* [78]. Additionally, uNK cell-derived interferon-γ (IFN-γ) promotes the vascularisation of the decidua and remodelling of spiral arteries [79,80]. IFN-γ activates the transcription of genes that are necessary for the maintenance of decidual integrity and angiogenesis and thus supports an adequate supply of the embryo. In this context, it was shown that mice lacking uNK cells display decreased levels of IFN-γ and thus abnormalities in the vascularisation pattern [81].

DC are also important modulators of decidual vascularisation during early pregnancy [82-84]. The role of DC was shown in a mouse model of reduced vascular expansion and maturation, in which DC were transiently ablated [82]. The ablation of DC resulted in implantation failure and embryo resorption. To ensure a proper vascular maturation in the decidua, uterine DC provide the critical factors sFlt-1 and TGF-β1. In mice with ablated DC, the adoptive transfer of $CXCR4^+DC$ rescued decidual vascularisation and remodelling of spiral arteries [83]. $CXCR4^+DC$ restored angiogenic processes by providing pro-angiogenic factors such as VEGF.

VEGF is one of the most important angiogenic factors. In early mouse pregnancy, VEGF is expressed in uterine tissue compartments that display an increased angiogenesis such as the epithelial stromal cells [85]. VEGF has been shown to bind VEGF receptor 2 (VEGFR2) on endothelial cells of the uterine stroma and decidua, thus promoting the angiogenesis processes during implantation and decidualization. The levels of VEGFR2 increase during the first days of gestation and are high in stromal endothelial cells during implantation and decidualization. The importance of VEGFR2-mediated angiogenesis was demonstrated in early mouse pregnancy, where a single dose of the VEGFR2 neutralizing antibody (DC101) during the implantation period resulted in complete foetal loss due to a severely reduced decidual vascularisation, while the inhibition of VEGFR1 or VEGFR3 did not disrupt pregnancy [86].

Recently, the expression of VEGF and VEGFR2 was determined during the third trimester of human pregnancy [87]. Both factors were differentially expressed in the vessels, stroma and glands of the decidua. In the placenta, villous CTB expressed VEGF and VEGFR2 throughout all stages of pregnancy. These results implicate that the VEGF-mediated angiogenesis via its receptor VEGFR2 is essential for the implantation, decidualization, and placentation processes during human and mouse pregnancy.

Interestingly, gal-1 binds neuropilin-1 (NRP-1) via its CRD, thereby facilitating VEGFR2 phosphorylation and thus promoting the migration and adhesion of endothelial cells [88]. NRP-1 specifically binds human $VEGF_{165}$ and mouse $VEGF_{164}$ isoforms, respectively, thereby enhancing the binding of VEGF to VEGFR2 and thus promoting angiogenesis [85,89]. In mice, the expression of NRP-1 increases during gestation [85]. NRP-1 is expressed on endothelial cells of the uterine stroma and decidua. NRP-1 is also expressed in the human decidua and placenta throughout gestation [87]. However, the contribution of gal-1 to this pathway regarding pregnancy-associated angiogenic processes remains to be elucidated.

1.4.2 Pro-angiogenic functions of gal-1

The role of gal-1 in the promotion of angiogenesis has been well documented in many physiological conditions apart from pregnancy [90]. Gal-1 expression and secretion is induced under hypoxic conditions, and facilitates VEGFR2 signalling in endothelial cells by binding to NRP-1 [88]. The binding of gal-1 to endothelial cells enhances H-Ras signalling, which is an early marker of activation [91,92]. Also, gal-1 displays pro-angiogenic, growth factor properties as its secretion stimulates proliferation, migration and tube formation (formation of capillary-like structures with a lumen) of endothelial cells [88,91-94]. Moreover, pathological conditions such as cancer are characterised by an enhanced angiogenesis and increased endothelial gal-1 levels [95]. However, *Lgals1* deficient mice are viable and fertile and their endothelial cells do not display an aberrant vascular phenotype [96]. Nevertheless, these mice show a defective angiogenesis under pathological conditions [92,97,98]. For example, the tumour growth is reduced in *Lgals1* deficient mice when compared to their wild type counterparts [97].

The knockdown of gal-1 expression using different approaches, including RNA interference, blocking peptides or antibodies, and competing carbohydrates, inhibits endothelial cell function [99]. For instance, in zebrafish embryos, the vascular guidance is disturbed during vessel formation with the injection of specific gal-1 antisense oligonucleotides [97]. Both these oligonucleotides and a neutralising gal-1 monoclonal antibody impairs the migration, proliferation, invasion, and tube formation of human umbilical vein endothelial cells (HUVEC) [88]. Furthermore, the neutralising antibody reduces the tumour growth *in vivo* by inhibiting angiogenic processes [91].

In the last decades, there has been a wealth of research to design new drugs that block angiogenic processes and thus contribute to anti-cancer therapy [100]. In this context, the *de novo* designed synthetic peptide anginex was shown to exert anti-angiogenic functions via the prevention of endothelial cell migration, adhesion and growth *in vitro* and *in vivo* [101-103]. Gal-1 was identified as the molecular target of anginex and the specific binding inhibits its pro-angiogenic functions [92,97]. These studies identify gal-1 as an important molecule during angiogenesis, but its role in angiogenic processes during pregnancy remains to be established.

1.5 Pregnancy complications and galectin-1

The irregular expression of gal-1 is associated with pregnancy complications [104,105]. The most common complication during early pregnancy is **spontaneous abortion (SA)**. SA are caused by many reasons, which cannot all be identified. About 15 to 20% of all clinically recognised pregnancies end in a SA, while the risk decreases with gestational age [106]. More than half of the cases result from foetal chromosomal abnormalities [107]. Other factors include anatomic abnormalities of the female genital organs, immune and endocrine diseases, hereditary disorders, chemical factors, and psychological aspects like unusual stress in the mother [108]. Women with SA display reduced circulating gal-1 levels in the first trimester of pregnancy, before the miscarriage occurs [41]. Likewise, gal-1 was reduced in the decidua and placental villi obtained from miscarriage patients when compared to healthy pregnant women [41,104]. Gal-1 is more sensitive than the typically used parameters (hCG and sonograph) for the prognosis and diagnosis of SA [41].

Recurrent spontaneous abortions (RSA) affect approximately 1% of women, and are defined as at least three consecutive miscarriages during attempts to conceive [109]. Similarly to SA patients, circulating levels of gal-1 are reduced in women suffering from RSA [46]. Ramhorst *et al.* also reported that these patients have a higher prevalence of gal-1 autoantibodies [46]. The analysis of gal-1 autoantibodies in serum of healthy pregnant women in our lab showed an increase during pregnancy progression [110]. Moreover, we did not observe significant differences in autoantibody levels between healthy pregnant and SA serum. Whether gal-1 autoantibodies are a cause or a consequence of RSA remains to be determined.

Another common pregnancy complication is **preeclampsia (PE)**, a multisystemic maternal syndrome that affects 12 to 22% of pregnancies. PE is the leading cause of perinatal mortality and morbidity [111,112]. In the United States, the analysis of pregnancy-related deaths over a four-year period revealed that PE directly caused 17.6% of all maternal deaths [113]. Since the clinical symptoms such as proteinuria, hypertension and endothelial dysfunctions appear in different severities and resolve after delivery, it is hypothesized that the placenta plays an essential role in PE development.

It has been proposed that PE should be defined in two classes: early onset (< 34 weeks of gestation) and late onset (> 34 weeks) PE, whereas mixed forms are very common [114]. The main difference is the severity of the clinical symptoms. In late onset PE, an abnormal maternal response towards the normally developed placenta occurs [115]. Women suffering from late onset PE carry predisposing factors such as hypertension, diabetes, and obesity, but the symptoms are rather mild. In contrast, early onset PE is characterised by a severe course of the disease and caused by a poor placentation during early pregnancy with an impaired remodelling of maternal spiral arteries resulting in increased hypoxia and thus oxidative stress (Figure 5) [116]. Due to the impaired utero-placental blood flow, the foetuses suffer from intrauterine growth restriction (IUGR) [117]. However, the classification of PE subtypes is still under discussion.

Generally, PE can be divided in pre-clinical and clinical stages that are defined by different foetal-maternal interfaces challenging maternal immune cells [116]. In the first pre-clinical stage, the blastocyst implants into the uterus, so that syncytio- and cytotrophoblasts (STB and CTB) come into direct contact with the maternal decidua. The second pre-clinical stage takes places during the placentation process and is characterized by invading extravillous

Introduction

CTB within the decidua and villous STB bathed in maternal blood in the intervillous space. During the first two stages the maternal immune system recognizes paternal alloantigens, but it is symptomless. Finally, the third stage is clinical and occurs after placentation in the second half of pregnancy, in which CTB still contact the decidua and villous STB are bathed in maternal blood.

The placenta may release factors resulting in an increased inflammatory response of the maternal immune system that causes endothelial injury and manifests in hypertension and proteinuria during PE [114]. These bioactive factors are excessively secreted into the maternal circulation by STB and include sFlt-1, soluble endoglin (sEng), corticotrophin-releasing hormone, activin-A, inhibin-A, leptin, and haem [118-124]. Moreover, the secretion of placental growth factor (PGF) is reduced from STB of preeclamptic women and thus compromises the placental function [125]. Besides the secretion of placental factors, micro- and nanovesicles are shed from the syncytial surface of the placenta and enter the maternal circulation during normal pregnancy [116,126]. Microvesicles activate the production of inflammatory cytokines and are increased in the circulation of preeclamptic women, thereby causing endothelial cell damage [127,128].

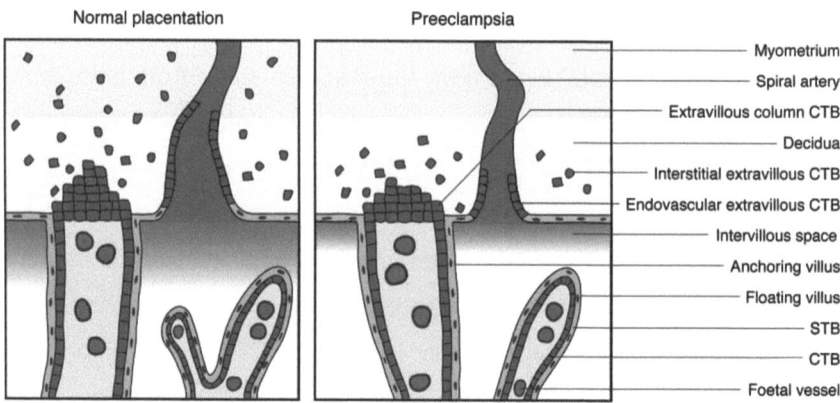

Figure 5: Poor placentation during preeclampsia

Introduction

In a healthy pregnancy, interstitial extravillous cytotrophoblasts (CTB) invade the maternal decidua and myometrium. In PE patients, the invasion is limited to the myometrium. Normally, the endovascular extravillous CTB replace the endothelial and smooth muscle cells of maternal spiral arteries and thus transform them into dilated blood vessels and ensure efficient blood flow in the intervillous space. During PE, the endovascular invasion is reduced and the spiral artery remodelling is compromised. As a consequence, PE placentas are characterised by a lower blood supply to the intervillous space, so that less nutrients and oxygen are taken up by the syncytiotrophoblasts (STB) on the chorionic villi and the foetus suffers from a reduced supply.

Gal-1 is up-regulated on extravillous CTB and stroma cells of the placenta and decidua in preeclamptic patients compared to healthy pregnant women [105,129]. Since hypoxia induces the expression of gal-1 at the mRNA and protein level, placental hypoxia, which is characteristic for PE, could explain the local increase of gal-1 in PE patients [130-132]. The numbers of gal-1-expressing natural killer (NK) and T cells are reduced in the periphery in PE patients, possibly leading to immune cell activation during PE pathogenesis [133]. To further this research prospective studies are needed, in which early and late onset PE are classified. Local and peripheral gal-1 expression should be determined during all stages of pregnancy. These studies have the potential to define the gal-1 contribution to PE pathogenesis and identify a possible application for gal-1 as a pre-clinical biomarker.

Gestational diabetes mellitus (GDM) is a pregnancy complication that affects 5-10% of pregnant women, increasing in prevalence and associated with increased maternal and infant morbidity and mortality during gestation and in later life [134-136]. GDM is defined as any degree of glucose intolerance that first appears during pregnancy and resolves after delivery [137]. Syncytiotrophoblasts in the placenta take up glucose from the maternal circulation via the glucose transporter (GLUT)-1 [138,139]. Interestingly, GLUT-1 expression on mouse embryonic stem cells is up-regulated by gal-1 [140]. Gal-1 is increased in third trimester placentas from GDM patients and could cause GLUT-1 up-regulation thus promoting an increased glucose transfer (*unpublished data*). The higher glucose levels in the foetus lead to the increased production of insulin and result in enhanced foetal growth.

Introduction

1.6 Aim and main results of the thesis

The aim of this thesis is to investigate the contribution of gal-1 to the angiogenic processes taking place at the foetal-maternal interface during pregnancy. We used the CD11c.DTR mouse to examine the effect of gal-1 supplementation in an established model of reduced vascular expansion [82]. Exogenous gal-1 supported implantation and placentation in these mice by increasing angiogenic factors and maintaining normal vascular development. Furthermore, we showed that gal-1 mediates its signals via VEGFR2, since pregnancy-protective effects were abolished when this pathway was blocked with a neutralising antibody [86,88]. Next, we investigated the influence on gal-1 inhibition during implantation and placentation in a normal mouse pregnancy using the synthetic peptide anginex [97]. Abrogation of gal-1–mediated angiogenesis caused a PE-like syndrome with a poor placentation, hypertension, proteinuria, and IUGR foetuses thus resembling the main PE symptoms in humans [117]. Completing these results, we describe the development of a late gestation PE-like syndrome in *Lgals1* deficient mice.

Adequate trophoblast cell functions are indispensable for the placentation process. During PE, extravillous cytotrophoblasts are less efficient in the remodelling of spiral arteries and contribute to a poor placental development [114]. To examine trophoblast functions under gal-1 inhibition, we used a human cell line treated with anginex. We found impaired branching and adherent capacity, which are indicative of a reduced capability for spiral artery remodelling. Finally, using two prospective studies, we assessed circulating and local levels of human gal-1 in samples from healthy pregnant women as well as women with early or late onset PE. Gal-1 levels were differentially regulated in the placenta and periphery of early and late PE patients when compared to healthy women. This thesis thus provides evidence for gal-1 involvement in angiogenic processes during pregnancy and PE pathogenesis and emphasises the need to distinguish between the two classes of PE. Gal-1 could be developed into a valuable biomarker for early PE development.

2. Methods

2.1 Laboratory equipment and consumables

96-well black/clear Imaging Plates	BD Falcon, Franklin Lakes, USA
96-well Plates	BD Falcon, Franklin Lakes, USA
Amersham Hypercassette autoradiography cassette	GE Healthcare, Buckinghamshire, UK
Amersham Hyperfilm ECL flim	GE Healthcare, Buckinghamshire, UK
Analysis balance ED124S	Sartorius, Goettingen, Germany
Axio Vert.A1 inverted phase-contrast microscope	Zeiss, Jena, Germany
Axiophot light microscope & AxioCam HRc camera	Zeiss, Jena, Germany
Balance LC6200D	Sartorius, Goettingen, Germany
BioPhotometer plus	Eppendorf, Hamburg, Germany
BZ9000 fluorescence microscope	Keyence, Offenbach, Germany
Cell Culture Flasks	BD Falcon, Franklin Lakes, USA
Centrifuge 5415C	Eppendorf, Hamburg, Germany
Centrifuge Heraeus® Labofuge® 400R	Sigma-Aldrich, Munich, Germany
CO_2 Incubator	Thermo Scientific, Braunschweig, Germany
CODA Non-Invasive Blood Pressure System	Kent Scientific Corporation, Torrington, USA

Methods

Consort EV202 power supply	Sigma-Aldrich, Munich, Germany
Cordless pestle motor	Sigma-Aldrich, Munich, Germany
Cover slips	R. Langenbrinck, Emmendingen, Germany
CryoPure tubes	Sarstedt, Nuembrecht, Germany
CryoStar NX70 cryostat	Thermo Scientific, Braunschweig, Germany
FLUOstar OPTIMA plate reader	BMG LABTECH, Ortenberg, Germany
Glassware	Schott, Mainz, Germany
Magnetic stirrer MR 3001	Heidolph, Kelheim, Germany
Manual cell counter	Carl Roth, Karlsruhe, Germany
Mastercycler nexus gradient	Eppendorf, Hamburg, Germany
Metabolic cages	FMI, Seeheim-Ober Beerbach, Germany
Micro tubes	Sarstedt, Nuembrecht, Germany
MicroAmp® Optical 96-Well Reaction Plate	Applied Biosystems, Foster City, USA
Microm HM 355 microtome	Thermo Scientific, Braunschweig, Germany
Microscope slides	Engelbrecht, Edermuende, Germany
Mini-PROTEAN® 3 Cell & Mini Trans-Blot Module	Bio-Rad, Munich, Germany
Mini-PROTEAN® Tetra Electrophoresis System	Bio-Rad, Munich, Germany
NanoDrop™ ND2000 UV/Vis	Thermo Scientific,

spectrophotometer	Braunschweig, Germany
Nitrocellulose Membrane (162-0115)	Bio-Rad, Munich, Germany
Nunc-Immuno™ Plates	Sigma-Aldrich, Munich, Germany
PAP-Pen	Dako, Glostrup, Denmark
Pellet pestles	Sigma-Aldrich, Munich, Germany
pH meter WTW Microprocessor	WTW, Weilheim, Germany
Pipette tips	Sarstedt, Nuembrecht, Germany
Pipettes Research	Eppendorf, Hamburg, Germany
PROTEC developing machine Compact 2 for	Siemens Healthcare, Erlangen, Germany
Rotilabo® embedding cassettes	Carl Roth, Karlsruhe, Germany
Shaker HS 250 basic	IKA Labortechnik, Staufen, Germany
SuperFrost® Plus microscope slides	R. Langenbrinck, Emmendingen, Germany
Surgical dissecting instruments	Carl Roth, Karlsruhe, Germany
TaqMan 7500 Fast Real-Time PCR System	Applied Biosystems, Foster City, USA
TissueRuptor	Qiagen, Hilden, Germany
Trial Kits Spectra/Por® 1 – MWCO 6,000-8,000 dialysis tube (4570)	Carl Roth, Karlsruhe, Germany
Tubes	BD Falcon, Franklin Lakes, USA
Vortex-Genie 2	Scientific Industries, Bohemia, USA
µ-Slides	Ibidi, Planegg, Germany

2.2 Reagents

Acetic acid (338826)	Sigma-Aldrich, Munich, Germany
Acetone (107021)	Merck, Darmstadt, Germany
Acid fuchsin (105231)	Merck, Darmstadt, Germany
Acrylamide/Bis Solution 30% (161-0158)	Bio-Rad, Munich, Germany
Aluminium potassium sulphate dodecahydrate (31242)	Sigma-Aldrich, Munich, Germany
Aminosilane (440140)	Sigma-Aldrich, Munich, Germany
Ammonium persulphate (APS) (17-1311-01)	GE Healthcare, Buckinghamshire, UK
Ammonium sulphate (A4418)	Sigma-Aldrich, Munich, Germany
Angiotensin II (A9525)	Sigma-Aldrich, Munich, Germany
Aprotinin (A3886)	Sigma-Aldrich, Munich, Germany
Azophloxine (11640)	Sigma-Aldrich, Munich, Germany
Bouin's solution (1012)	Dr. K. Hollborn & Soehne, Leipzig, Germany
BSA (A7030)	Sigma-Aldrich, Munich, Germany
CellTracker™ green CMFDA (C2925)	Invitrogen, Carlsbad, USA
Chloral hydrate (15307)	Sigma-Aldrich, Munich, Germany
Chloroform (C7559)	Sigma-Aldrich, Munich, Germany
Citric acid (C0759)	Sigma-Aldrich, Munich, Germany
DAB substrate (K3467)	Dako, Glostrup, Denmark

Methods

DAPI stock (10236276001)	Roche, Indianapolis, USA
DEPC (159220)	Sigma-Aldrich, Munich, Germany
Diphtheria toxin (DT) (D0564)	Sigma-Aldrich, Munich, Germany
DMSO (20385)	Serva, Heidelberg, Germany
DNase I, amplification grade (1 U/µl) DNase I Reaction Buffer (10x) EDTA (25 mM, pH 8.0) (18060-015)	Invitrogen, Carlsbad, USA
dNTP mix (100 mM each) (10297-018)	Invitrogen, Carlsbad, USA
Dual Endogenous Enzyme Block (52003)	Dako, Glostrup, Denmark
ECL Western Blotting System (RPN2108)	GE Healthcare, Buckinghamshire, UK
EDTA solution (500 mM, pH 8.0) (324504)	Merck, Darmstadt, Germany
Eosin (2C-284)	Waldeck, Muenster, Germany
Ethanol, absolute (107017)	Sigma-Aldrich, Munich, Germany
Ethanol, methylated (216925)	Herbeta Arzneimittel, Berlin, Germany
FCS (S0615)	Biochrom, Berlin, Germany
Formalin (104002)	Merck, Darmstadt, Germany
Galectin-1 (gal-1) human, recombinant (G7420)	Sigma-Aldrich, Munich, Germany
Glacial acetic acid (101830)	Merck, Darmstadt, Germany
Glycine (3908)	Carl Roth, Karlsruhe, Germany
Hematoxylin (H3136)	Sigma-Aldrich, Munich, Germany
Hydrogen peroxide (H_2O_2) (107210)	Merck, Darmstadt, Germany
Isopropanol (I9516)	Sigma-Aldrich, Munich, Germany

KCl (6781)	Carl Roth, Karlsruhe, Germany
L-Glutamine (G6392)	Sigma-Aldrich, Munich, Germany
Laemmli Sample Buffer (4x) (161-0747)	Bio-Rad, Munich, Germany
Leupeptin (62070)	Sigma-Aldrich, Munich, Germany
Light Green (HT10316)	Sigma-Aldrich, Munich, Germany
Losartan (Y0001076)	Sigma-Aldrich, Munich, Germany
Matrigel™ (356230)	BD Biosciences, San Jose, USA
Methanol (107018)	Merck, Darmstadt, Germany
Milk powder (T145)	Carl Roth, Karlsruhe, Germany
NaCl (3957)	Carl Roth, Karlsruhe, Germany
NaOH (106498)	Merck, Darmstadt, Germany
Neo-Clear (109843)	Merck, Darmstadt, Germany
Normal donkey serum (017-000-001)	Jackson ImmunoResearch, West Grove, USA
Normal goat serum (005-000-001)	Jackson ImmunoResearch, West Grove, USA
Nutrient Mixture F-10 Ham (N6013)	Sigma-Aldrich, Munich, Germany
Orange G (115925)	Merck, Darmstadt, Germany
Paraffin Paraplast Plus	Leica Biosystems, Richmond, USA
PBS, sterile filtered (H15-002)	PAA Laboratories, Pasching, Austria
Penicillin-Streptomycin (P4333)	Sigma-Aldrich, Munich, Germany

Pepstatin (77170)	Sigma-Aldrich, Munich, Germany
Periodic acid (77310)	Sigma-Aldrich, Munich, Germany
Phosphomolybdic acid (100532)	Merck, Darmstadt, Germany
Ponceau S (78376)	Sigma-Aldrich, Munich, Germany
Ponceau xylidine (P2395)	Sigma-Aldrich, Munich, Germany
Power SYBR® Green PCR master mix (4368577)	Applied Biosystems, Foster City, USA
Precision Plus Protein Dual Xtra Standard (161-0377)	Bio-Rad, Munich, Germany
Protein Assay Dye Reagent Concentrate (500-0006)	Bio-Rad, Munich, Germany
Protein-Block (PHA-70874)	Dianova, Hamburg, Germany
Random Primers (3 µg/µl) (48190-011)	Invitrogen, Carlsbad, USA
ReBlot Plus Mild Antibody Stripping Solution (2502)	Merck, Darmstadt, Germany
RNaseOUT™ Recombinant Ribonuclease Inhibitor (40 U/µl) (10777-019)	Invitrogen, Carlsbad, USA
Schiff reagent (X900)	Carl Roth, Karlsruhe, Germany
Shandon Immu-Mount™ (99-904-02)	Thermo Scientific, Braunschweig, Germany
Sodium bisulphite solution ($NaHSO_3$) (13438)	Sigma-Aldrich, Munich, Germany
Sodium citrate (71497)	Sigma-Aldrich, Munich, Germany
Sodium dodecyl sulphate (SDS) (2326)	Carl Roth, Karlsruhe, Germany
Sodium iodate (S4007)	Sigma-Aldrich, Munich,

Methods

	Germany
Sodium phosphate monobasic monohydrate (NaH$_2$PO$_4$ x H$_2$O) (71504)	Sigma-Aldrich, Munich, Germany
Staurosporine (S4400)	Sigma-Aldrich, Munich, Germany
Sulphuric acid (H$_2$SO$_4$) (339741)	Sigma-Aldrich, Munich, Germany
SuperScript® II Reverse Transcriptase (200 U/µl) First-Strand Buffer (5x) DTT (0.1 M) (18064-014)	Invitrogen, Carlsbad, USA
TEMED (T9281)	Sigma-Aldrich, Munich, Germany
Tissue Freezing Medium (0201 08926)	Jung, Nussloch, Germany
TMB (87748)	Sigma-Aldrich, Munich, Germany
Tris (45-000-236)	GE Healthcare, Buckinghamshire, UK
Tris base (648310)	Merck, Darmstadt, Germany
Tris HCl (648317)	Merck, Darmstadt, Germany
Triton X-100 (X100)	Sigma-Aldrich, Munich, Germany
Trypsin (T9935)	Sigma-Aldrich, Munich, Germany
Trypsin-EDTA 0.25% (25200)	Invitrogen, Carlsbad, USA
Tween-20 (P1379)	Sigma-Aldrich, Munich, Germany
VEGFR2 antibody (clone DC101) (BE0060)	BioXCell, West Lebanon, USA
Vitro-Clud® (04-001)	R. Langenbrinck, Emmendingen, Germany
Weigert's iron hematoxylin kit (1159973)	Merck, Darmstadt, Germany
Xylene (108661)	Merck, Darmstadt, Germany

| α-Amylase (10070) | Sigma-Aldrich, Munich, Germany |
| β-Mercaptoethanol (M3148) | Sigma-Aldrich, Munich, Germany |

2.3 Buffers and solutions

Cell culture medium (*in vitro* assays)	Ham's F10 with 10% FCS, 100 U/ml Penicillin, 100 µg/ml Streptomycin, 2 mM glutamine
Citrate buffer	18 ml 0.1 M citric acid + 82 ml 0.1 M sodium citrate + 900 ml Aqua dest.; pH 6.0
DAPI solution	100 µl DAPI stock (2 mg/ml) + 100 ml Methanol
Hematoxylin solution	1 g hematoxylin + 0.2 g sodium iodate + 50 g aluminium potassium sulphate dodecahydrate + 50 g chloral hydrate + 1 citric acid ad 1 l with Aqua dest.
PBS (10x)	80 g NaCl + 2 g KCl + 28 g $NaH_2PO_4 \times H_2O$ + 4 g NaOH (pH 6.81) ad 1 l with Aqua dest.
Ponceau solution	0.5% Ponceau S + 1% acetic acid in Aqua dest.
TAE buffer (50x)	242 g Tris base + 57.1 ml glacial acetic acid + 100 ml EDTA solution ad 1 l with Aqua dest.
TBS	0.9 g Tris base + 6.85 g Tris HCl + 8.78 g NaCl (pH 7.5) ad 1 l with Aqua dest.
TMB substrate	0.24 g TMB + 5 ml EtOH (absolute) + 5 ml DMSO 1:100 in 0.2 M citric acid (pH 3.95) + H_2O_2 (3.4 µl per 10 ml citric acid)
Western blot running buffer (10x)	1.92 M glycine + 1% SDS + 0.25 M Tris ad 1 l with Aqua dest.
Western blot transfer	192 mM glycine + 25 mM Tris ad 1 l with Aqua dest.

buffer (10x)   Dilute 1:10 with 20% methanol for 1x transfer buffer

2.4 Antibodies

Primary Antibody	Company	Catalogue number
Biotin-conjugated Dolichos biflorus agglutinin (**DBA**)	Sigma-Aldrich	L6533
Biotin-conjugated Isolectin B4 (**IB4**)	Sigma-Aldrich	L2140
rat monoclonal anti-mouse **endoglin** IgG$_{2a}$ (MJ7/18)	Santa Cruz Biotechnology	sc-18893
rat monoclonal anti-mouse **CD31** / PECAM-1 IgG$_{2a}$ (MEC 13.3)	Santa Cruz Biotechnology	sc-18916
rabbit polyclonal anti-mouse/human **galectin-1** (gal-1) IgG (H-45)	Santa Cruz Biotechnology	sc-28248
goat polyclonal anti-mouse **placental lactogen-I** (PL-I) IgG	Santa Cruz Biotechnology	sc-34713
goat polyclonal anti-mouse **proliferin** (PLF) IgG	Santa Cruz Biotechnology	sc-47345
HRP-conjugated anti-**β-actin**	Sigma-Aldrich	A3854

Secondary Antibody	Company	Catalogue number
ExtrAvidin®-**Peroxidase**	Sigma-Aldrich	E2886
Rhodamine (**TRITC**) AffiniPure goat anti-rat IgG	Jackson ImmunoResearch	112-025-167
Rhodamine (**TRITC**) AffiniPure F(ab')2 Fragment goat anti-rabbit IgG	Jackson ImmunoResearch	111-026-045
Rhodamine (**TRITC**)	Jackson	705-025-003

| AffiniPure donkey anti-goat IgG | ImmunoResearch | |
| HRP-conjugated anti-rabbit IgG | Sigma-Aldrich | A0545 |

2.5 Animal models

All mice were maintained in the Charité animal facility (Forschungseinrichtung für experimentelle Medizin, Charité Campus Virchow-Klinikum) with a 12 h light-dark cycle and feeding *ad libitum*. Five- to six-week old female mice were mated with at least 8-week old males. The presence of a vaginal plug after cohabitation was denoted as gestation day (gd) 0.5. All intraperitoneal (i.p.) injections were prepared in a total volume of 200 µl with sterile PBS. In all animal experiments, 6-8 mice per group were analysed.

CD11c.DTR transgenic mice (B6.FVB-Tg(Itgax-DTR/GFP)57Lan/J; JAX© Mice) express the diphtheria toxin receptor (DTR) under the control of the *CD11c* promoter [141]. Balb/c (JAX© Mice)-mated CD11c.DTR female mice received i.p. injections according to Table 2. The human recombinant galectin-1 (gal-1) supplemented gal-1 levels exogenously. The Diphtheria toxin (DT) was used to ablate $CD11c^+$ dendritic cells (DC), i.e. approximately 8 h after the DT injection all $CD11c^+$ DC were depleted, while the population gradually recovered after two days. The neutralizing VEGFR2 antibody (clone DC101) prevented gal-1 signalling via the VEGFR2 pathway. Pregnancy fitness was calculated for selected groups as follows: pregnancy fitness [%] = (pregnant females x 100) / all females with vaginal plug.

Table 2: Treatment of CD11c.DTR female mice
CD11c.DTR females with a vaginal plug (denoted as gd 0.5) were treated with different combinations of the substances listed in the table. The given numbers specify the gestation day(s) of treatment. — means no treatment with the respective substance.

Treatment per mouse and gd	Group			
	gal-1	aDC	aDC + gal-1	aDC + gal-1 + DC101

10 µg galectin-1 human, recombinant	4.5-6.5	—	4.5-6.5	4.5-6.5
4 ng/g body weight diphtheria toxin	—	4.5	4.5	4.5
1.32 mg anti-VEGFR2 (DC101)	—	—	—	4.5-6.5

C57BL/6J (JAX© Mice) female mice were mated with Balb/c males and received i.p. injections of 0.5 mg/kg body weight of the synthetic peptide anginex (kindly provided by Dr Victor L.J.L. Thijssen, Angiogenesis Laboratory, VU University Medical Center, Amsterdam, Netherlands) from gd 4.5 to 12.5. Anginex inhibits the angiogenic function of gal-1 [97,101]. Control C57BL/6 females received sterile PBS.

129/P3J gal-1$^{-/-}$ (***Lgals1* ko/deficient**) and 129/P3J gal-1$^{+/+}$ (***Lgals1* wt**) female and male mice (Institut Jacques Monod, Paris) were mated without any additional injections. Six *Lgals1* ko females received 0.5 mg/kg body weight anginex.

2.6 Experimental sampling

The mice in the respective groups were anesthetized on different gd and blood was taken by retro-orbital puncture for serum separation by centrifugation. After killing the mice by cervical dislocation, all implantations were photo-documented and the implantation failure was calculated at gd 7.5: implantation failure = (non-viable implantations x 100) / total number of implantations. The uteri were embedded in Tissue Freezing Medium and frozen on dry ice for cryo sectioning, or formalin-fixed (5% in PBS) for paraffin sections. Some uteri were frozen in liquid nitrogen for isolation of total protein. At later gd, the foetal loss and pregnancy success was calculated, respectively. Foetal loss = (foetal resorptions x 100) / total number of implantations; pregnancy success = 100 − ((foetal resorptions x 100) / total number of implantations). Whole implantations were formalin-fixed for paraffin sections or frozen for cryostat sections. Some implantations were separated into decidua, placenta and embryo. The placental weight was noted and the placenta and decidua were frozen in liquid nitrogen for isolation of total protein or RNA. The embryos were fixed in Bouin's solution and subsequently cleared in 70% EtOH for photo documentation, body weight measurement and Theiler stage analysis (Table 11 in appendix) [142].

2.7 Human samples

The human samples were kindly provided by Professor Dr Anne Cathrine Staff from Oslo University Hospital, Norway and are part of the research biobank collection at Oslo University Hospital approved by the Regional Committee of Medical Research Ethics in Eastern Norway. Women included in this study had singleton pregnancies, no pre-existing hypertension or other chronic diseases, no rupture of uterine membranes or clinical signs of infection. Samples from 36 women with uncomplicated, normotensive pregnancies (control) and 35 preeclamptic women (early and late PE) were included in this study.

Preeclampsia was defined by hypertension (> 140/90 mmHg) after 20 weeks of gestation on more than two occasions 6 h apart and proteinuria defined by protein dip stick ≥ 1+ on more than two midstream urine samples 6 hours apart or a urine excretion of ≥ 0.3 g protein within 24 h in the absence of a urinary infection [117]. Clinical characteristics are summarised in Table 3. All women underwent Caesarean section, in the control group due to breech presentation or other reasons. Blood was taken in the third trimester before the onset of labour. Circulating sFlt-1 and PGF levels were determined by ELISA (R&D Systems, DVR100B and DPG00) at Oslo University Hospital. Circulating gal-1 levels were measured as described below. Placental biopsies were obtained from central non-infarcted cotyledons and decidual tissue was collected by vacuum suctioning of the placental bed after gentle placental delivery [143].

Table 3: Clinical characteristics of the Oslo cohort

Methods

Parameter	Control (n = 36)	Early PE (n = 18)	Late PE (n = 17)
Age (years)	32.1 ± 4.9	32.7 ± 3.2	31.5 ± 5.5
GA [A] (weeks)	38.4 ± 0.6	30.2 ± 2.9	38.0 ± 0.8
Systole (mm Hg)	122.9 ± 16.0	174.1 ± 24.0	160.1 ± 16.2
Diastole (mm Hg)	71.3 ± 7.6	100.2 ± 10.4	98.2 ± 6.2
Placenta weight (g)	586.5 ± 129.0	243.8 ± 151.6	595.4 ± 201.8
sFlt-1 (pg/ml)	12,064.6 ± 5,344.5	18,866.7 ± 1,762.6	17,517.1 ± 2,625.3
PGF (pg/ml)	161.9 ± 65.1	48.4 ± 38.9	124.4 ± 57.7

[A] GA = gestational age at delivery

A prospective cohort study was conducted at the Department of Paediatrics, Women and Infants Hospital of Rhode Island, USA. Written informed consent was obtained from all women. The study was approved by the Women and Infants Hospital Institutional Review Board. Blood samples were collected during the second trimester and kindly provided by Professor Dr Surendra Sharma. Circulating gal-1 levels were determined by ELISA as described below. The clinical characteristics of the analysed study population are summarized in Table 4.

Table 4: Clinical characteristics of the Brown University cohort

Parameter	Control (n = 8)	Late PE (n = 8)
Age (years)	28.3 ± 4.5	29.9 ± 4.2
GA [A] (weeks)	39.8 ± 1.2	39.6 ± 1.9
Systole (mm Hg) 22 wk [B]	93.8 ± 5.2	103.8 ± 9.2
Diastole (mm Hg) 22 wk [B]	61.3 ± 3.5	67.5 ± 10.4
Systole (mm Hg) 32 wk [C]	96.3 ± 5.2	118.1 ± 4.7
Diastole (mm Hg) 32 wk [C]	61.2 ± 3.5	71.9 ± 2.5
Birth weight (kg)	2.9 ± 0.3	2.6 ± 0.7

[A] GA = gestational age at delivery;
[B] 22 wk = 22nd week of gestation;
[C] 32 wk = 32nd week of gestation

2.8 Immunohistochemistry and immunofluorescence

2.8.1 Hematoxylin & eosin (H&E) stain

Frozen mouse uteri or kidneys were sectioned at the cryostat (8 µm) and fixed in acetone for 10 min at -20 °C. The slides were dried for 10 min at RT, encircled with PAP-Pen and washed for 5 min in TBS. Cell nuclei were stained with filtered hematoxylin for 12 min. Sections were washed three times for 5 min with tap water. Other cell structures were counterstained with filtered eosin for 15 min. Sections were dehydrated in 100% EtOH (2 x 2 min), cleared in xylene (2 x 10 min) and mounted in Vitro-Clud®. Tissue sections were examined using a light microscope.

2.8.2 IB4 stain

Formalin-fixed uteri were dehydrated for paraffin embedding according the following protocol: 30 min 70% EtOH – 30 min 80% EtOH – 15 min 96% EtOH – 15 min 96% EtOH – 30 min 96% EtOH – 15 min 100% EtOH – 30 min 100% EtOH – 45 min Neo-Clear – 45 min Neo-Clear – 2 h paraffin (60 °C) – embedding – overnight -20 °C.

Pre-cooled (-20 °C) paraffin-embedded uteri were sectioned at 4 µm and mounted on aminosilane-coated slides. For staining, the slides were deparaffinised in xylene (2 x 10 min) and rehydrated in a descending alcohol series (2 min each step: 2 x 100% - 96% - 90% - 80% - 70% EtOH). The sections were washed in Aqua dest. (2 x 5 min) and PBS (2 x 5 min) and cooked for 10 min in a pressure cooker with citrate buffer. After washing for 5 min in PBS, sections were encircled with PAP-Pen. Endogenous peroxidase activity was blocked with Dual Endogenous Enzyme Block for 30 min at RT. After washing with PBS (3 x 5 min), sections were blocked with Protein-Block for 20 min at RT and incubated with biotin-conjugated IB4 (1:200 in PBS) for 2 h at RT. The sections were washed, incubated with ExtrAvidin®-Peroxidase (1:200 in PBS) for 30 min at RT and washed again. The DAB substrate was incubated for 6 min at RT for colour reaction. After washing three times 10 min in PBS, nuclei were counterstained in filtered hematoxylin for 1 min 30 s and washed again for 5 min in tap water. The sections were dehydrated in an ascending alcohol series (2 min each step: 70% - 80% - 90% - 96% - 2 x 100% EtOH), cleared in xylene (2 x 10 min) and mounted in Vitro-Clud®. Tissue sections were examined using the light microscope and analysed with the AxioVision software (Zeiss, Germany).

2.8.3 DBA-PAS double stain

Uteri were sectioned, deparaffinised, dehydrated and cooked as described in the previous section. Samples were incubated 20 min with 1% α-Amylase and washed three times in PBS. Endogenous peroxidase activity was blocked with Dual Endogenous Enzyme Block for 30 min at RT. After washing with PBS (3 x 5 min), sections were blocked with Protein-Block for 20 min at RT and incubated with biotin-conjugated DBA (1:2000 in PBS) overnight at 4 °C. The sections were washed, incubated with ExtrAvidin®-Peroxidase (1:200 in PBS) for 30 min at RT and washed again. The DAB substrate was incubated for 7 min at RT. After washing three times for 10 min in PBS, sections were incubated in 1% periodic acid for 20 min at RT and washed 5 min in tap water. The Schiff reagent was incubated 35 min at RT and sections were washed 5 min in tap water. Sections were incubated 0.5% $NaHSO_3$ and washed (5 min tap water). Nuclei were counterstained with filtered hematoxylin for 1 min 30 s. Sections were washed 5 min in tap water, dehydrated and mounted. DBA^+PAS^+ cells were counted with a cell counter.

2.8.4 Periodic acid-Schiff (PAS) stain

Sections were deparaffinised and rehydrated as described previously. Periodic acid (1% in Aqua dest.) was incubated for 10 min at RT. Sections were washed three times for 10 min with tap water and incubated 1 min in 70% EtOH. Schiff reagent was incubated 5 min at RT and sections were washed (3 x 10 min tap water). Nuclei were counterstained with filtered hematoxylin for 1 min and washed 5 min in tap water. The sections were dehydrated and mounted as described above. PAS^+ cells in the placenta were counted with a cell counter.

2.8.5 Masson-Goldner's trichrome stain

Serial paraffin-embedded uterine sections from gd 10.5 or 13.5 were deparaffinised and rehydrated as described above. Nuclei were counterstained with Weigert's iron hematoxylin for 3 min. Slides were then stained with Solution A (azophloxine, acid fuchsin, and Ponceau xylidine in 0.2% acetic acid) for 7 min and rinsed in 1% acetic acid followed by Solution B (phosphomolybdic acid-Orange G solution) for 2.5 hours and rinsed again with 1% acetic acid. In the last step, slides were incubated with Solution C (Light Green in 0.2% acetic acid) for 10 min followed by dehydration and mounting as described above. Tissue sections were examined using a light microscope and spiral artery walls were analysed with the AxioVision software (Zeiss, Germany).

2.8.6 Immunofluorescence stainings

Mouse uteri or human placenta sections were dried for 10 min at RT, encircled with PAP-Pen, washed in TBS (5 min) and blocked with 2% normal serum for 20 min. Primary antibodies are listed above and were incubated in the following dilutions overnight at 4 °C: **endoglin, PLF, gal-1** 1:200 and **CD31, PL-I** 1:100 in TBS. In negative controls, the primary antibodies were replaced by irrelevant IgG. After washing with TBS, stained sections were incubated 1 h at RT with TRITC-conjugated secondary antibodies (listed above, dilutions: endoglin, PL-I, PLF, gal-1 1:200; CD31, 1:100 in TBS). Nuclei in all sections were counterstained by incubating 5 min in DAPI solution, followed by washing with TBS and mounting in Shandon Immu-Mount™. Sections were analysed with the fluorescence microscope.

2.8.7 Detection of hypoxia

Hypoxic cells in the placenta of mice were detected with intravenously injected pimonidazole-HCl (50 mg/kg body weight; Hypoxyprobe, Inc., Hypoxyprobe™-1 Plus Kit, HP2-100) 60 min before killing. Cryo sections (8 µm) of uteri from gd 10.5 were washed in TBS and blocked with Protein-Block for 20 min. Slides were then incubated 1 h at RT with the FITC-conjugated monoclonal antibody (1:100 in TBS, provided with the kit). Nuclei were counterstained by incubating 5 min in DAPI solution. Slides were washed and mounted in Shandon Immu-Mount™. Sections were examined with the fluorescence microscope. Mean grey values were measured with ImageJ software (NIH, USA).

2.9 Protein isolation and Western blot analysis

Protein from human placenta was isolated with the Total Protein Extraction Kit (Chemicon (Millipore), 2140). All steps were performed on ice or at 4 °C. 1 g of tissue was incubated 5 min in 2.5 ml 1x PI (protease inhibitors diluted in TM Buffer) and homogenized three times for 20 s with a pellet pestle. The lysate was centrifuged for 20 min at 16,000 x g. The supernatant contained the total isolated protein.

Protein concentration was determined with a Bradford assay. BSA standards were prepared from 10 to 0.1563 mg/ml. The isolated protein (1 µl) or BSA standard was mixed with 250 µl of diluted Protein Assay Dye Reagent (1:4 with Aqua dest.) in a microtiter plate (Nunc-Immuno™ Plate, 456537) and incubated for 10 min at RT. The optical density was read at 595 nm.

Equal amounts of protein from human placenta samples (40 µg) were mixed with 5 µl Laemmli buffer and 1 µl β-ME (ad 20 µl with Aqua dest.) and resolved by SDS-PAGE (Table 5). The gel electrophoresis was performed with the Mini-PROTEAN® Electrophoresis System (Bio-Rad) in Western blot running buffer. The Precision Plus Protein standard (7.5 µl) was loaded as a marker. The electrophoresis was performed at 100 V for 10 min and 160 V for approximately 80 min.

Table 5: 15% polyacrylamide gel for the resolution of total isolated protein

	Separation gel		Stacking gel	
Aqua dest.		2.3 ml		2.1 ml
Acrylamide/Bis solution		5.0 ml		500 µl
Tris	1.5 M (pH 8.8)	2.5 ml	1 M (pH 6.8)	380 µl

10% SDS	100 µl	30 µl
10% APS	100 µl	30 µl
TEMED	4 µl	3 µl

Proteins were transferred to nitrocellulose membranes with the Mini-PROTEAN® 3 Cell & Mini Trans-Blot Module (Bio-Rad) in Western blot transfer buffer for 60 min at 340 mA. Equal loading was confirmed with Ponceau solution.

Membranes were blocked with 5% milk and 0.1% Tween-20 in PBS for 1 h at RT and incubated with the gal-1 antibody (1:1500 in 3% BSA and PBS; Santa Cruz Biotechnology, sc-28248) overnight at 4 °C. The membranes were washed with 5% milk and 0.1% Tween-20 in PBS (2 x 5 min and 2 x 15 min). The HRP-conjugated antibody (1:5000 in 3% BSA and PBS; Sigma-Aldrich, A0545) was incubated for 1 h at RT and membranes were washed with 5% milk and 0.1% Tween-20 in PBS (2 x 5 min and 1 x 15 min) and 15 min with PBS. Blots were developed using the enhanced chemiluminescence (ECL) reagent and films in a developing machine. The antibodies were stripped with ReBlot Stripping Solution for 30 min at RT. Membranes were washed with PBS and the HRP-conjugated anti-β actin antibody (1:50,000 in 1% BSA and PBS; Sigma-Aldrich, A3854) was incubated 1 h at RT. Blots were developed and proteins were semi-quantified using ImageJ software (NIH, USA). Gal-1 protein levels are expressed relative to β-actin for each sample.

2.10 Angiogenesis array

We used the Mouse Angiogenesis Array Kit (R&D Systems, ARY015) to determine the levels of angiogenesis-related proteins. Fifty-three different capture antibodies are printed on nitrocellulose membranes in duplicates. Tissue of whole mouse implantations was homogenized with protease inhibitors (10 µg/ml each) in PBS. After the addition of 1% Triton X-100, the lysates were frozen at -80 °C. The next day, the samples were thawed and centrifuged at 10,000 x g for 5 min to remove cellular debris. Protein concentrations were determined by Bradford assay as described previously. The nitrocellulose membranes were blocked for 1 h at RT before adding a pre-incubated (1 h, RT) cocktail of 300 µg of protein and detection antibody. After the overnight incubation at 4 °C on a shaker, the membranes were washed and a streptavidin-HRP antibody was incubated for 30 min at RT on a shaker. After washing, the Chemi-Reagent Mix was incubated for 1 min and the membranes were exposed to X-ray films for 2 min. For the analysis, the pixel density was measured in each spot of the array with ImageJ software (NIH, USA), the background signal was subtracted for every protein and the mean of the duplicates was calculated. The corresponding signals of two different membranes, i.e. two experimental groups are compared to determine the relative change in protein expression.

2.11 RNA isolation, RT-PCR and quantitative real time PCR

2.11.1 Mouse

Total RNA was isolated from mouse placenta with the RNeasy Protect Mini Kit (Qiagen, 74124). A maximum of 30 mg frozen placenta was homogenised in 350 µl Buffer RLT (contains β-ME) with a pellet pestle. After centrifugation 3 min at full speed (16,000 x g), the supernatant was mixed with 350 µl 70% EtOH for precipitating RNA, transferred to the RNeasy spin column and centrifuged for 15 s at full speed. Washing was performed with 700 µl Buffer RW1 and centrifugation for 15 s at full speed followed by 500 µl Buffer RPE and centrifugation. A last washing step was accomplished with 500 µl Buffer RPE and centrifugation for 2 min. To elute RNA, 30 µl RNase-free water was added and the column was centrifuged 1 min at full speed. This step was repeated with the flow-through.

The concentration and quality of the isolated RNA was determined for reverse transcription (RT)-PCR with a BioPhotometer. cDNA was generated from 1 µg of isolated RNA. DnaseI (0.5 U) digestion was performed for

15 min at RT and stopped with 2.5 mM EDTA at 65 °C for 15 min. The RNA was incubated with Random Primers (0.25 µg/µl) and dNTP mix (10 mM each) for 5 min at 65 °C. The RNA primer mixture was transcribed to cDNA with SuperScript II (200 U), 10 mM DTT and 40 U RNaseOUT at 25 °C for 10 min, 42 °C for 50 min and 70 °C for 15 min.

The quantitative real time PCR was performed with the TaqMan® (Applied Biosystems). Each reaction contained 1 µl cDNA, 6.25 µl *Power SYBR®* Green PCR master mix, 3.75 µl DEPC-treated water and 900 nM of the appropriate forward and reverse primers (TIB MOLBIOL, Germany; Table 6). The PCR program was as follows: 2 min at 50 °C, 10 min at 95 °C, 40 cycles of 15 s at 95 °C and 60 s at 60 °C. A subsequent melting curve analysis was performed: 70 cycles of 10 s with a temperature increment of 0.5 °C per cycle starting at 60 °C. We calculated the relative expression (RE) with the following equation:

$$RE = 2^{-\Delta Ct}, \text{ in which } Ct = Ct_{\text{gene of interest}} - Ct_{\text{reference gene}}.$$

Table 6: Mouse primer sequences for qPCR in placental tissue

Gene	Forward primer (5'-3')	Reverse primer (5'-3')
Renin	TCTGGGCACTCTTGTTGCTC	GGGGGAGGTAAGATTGGTCAA
AT1	TCGCTACCTGGCCATTGTC	TGACTTTGGCCACCAGCAT
VEGF	ATCTTCAAGCCGTCCTGTGT	GCATTCACATCTGCTGTGCT
HIF-2α	TGAGTTGGCTCATGAGTTGC	TATGTGTCCGAAGGAAGCTG
PGF	CCACGCTCCTGTGAAACTAGA	GACCAAACCTCAAAGCATGG
HPRT	GTTGGATACAGGCCAGACTTTGT	CACAGGACTAGAACACCTGC

2.11.2 Human

Total RNA from human placental and decidual tissue was isolated with a combined protocol of QIAzol® (Qiagen, 79306) and RNeasy Mini Kit (Qiagen, 74104) including the RNase-Free DNase Set (Qiagen, 79254). The tissue was placed into QIAzol Lysis Reagent (1 ml per 100 mg tissue) and processed with the TissueRuptor until the lysate was homogeneous. The lysate was kept at RT for 5 min before the addition of chloroform (0.2 ml per 1 ml QIAzol Lysis Reagent) and vigorous mixing for 15 s. After 3 min at RT, the lysate was centrifuged for 15 min at 4 °C and 12,000 x g. The upper, aqueous phase was mixed thoroughly with isopropanol (0.5 ml per 1 ml QIAzol Lysis Reagent) and incubated for 10 min at RT. After centrifugation (10 min at 4 °C and 12,000 x g), the supernatant was mixed with 75% EtOH (1 ml per 1 ml QIAzol Lysis Reagent) and centrifuged again (5 min at 4 °C and 7,500 x g). The pellet contained the isolated RNA, was air-dried briefly and dissolved in 100 µl RNase-free water. After the addition of 350 µl Buffer RLT (contains β-ME) and 250 µl 100% EtOH, the sample was transferred to the RNeasy spin column and centrifuged 15 s at full speed (16,000 x g). 350 µl Buffer RW1 was added and centrifuged again for 15 s at full speed. The DNase I mix (10 µl DNase I stock + 70 µl Buffer RDD) was incubated with the sample for 15 min at RT. Again, 350 µl Buffer RW1 was added and centrifuged for 15 s at full speed. The spin column containing the RNA was washed two times with 500 µl Buffer RPE (15 s and 2 min at full speed). The isolated RNA was eluted with 30 µl RNase-free water by centrifugation for 1 min at full speed.

The concentration and quality of the isolated RNA was determined with the NanoDrop™ UV/Vis spectrophotometer. With the Transcriptor First Strand cDNA Synthesis Kit (Roche, 04 379 012 001), 1 µg of isolated RNA was reverse transcribed to cDNA. RNA and Random Hexamer Primer were incubated 10 min at 65 °C. Transcriptor Reverse Transcriptase, Reaction Buffer, Protector RNase Inhibitor and dNTP Mix were added and incubated 10 min at 25 °C and 60 min at 50 °C. The Reverse Transcriptase reaction was inactivated for 5 min at 85 °C.

The quantitative real time PCR was performed with the TaqMan® (Applied Biosystems) as described for cDNA from mouse placenta in section

2.11.1. The primers are listed in Table 7. We calculated the expression of *Lgals1* in relation to the *18s* housekeeping gene.

Table 7: Human primer sequences for qPCR in placental and decidual tissue

Gene	Forward primer (5'-3')	Reverse primer (5'-3')
Lgals1	TCGCCAGCAACCTGAATCTC	GCACGAAGCTCTTAGCGTCA
18s	Eukaryotic 18S rRNA (GenBank accession number: X03205) (PE Biosystems)	

2.12 Enzyme-linked immunosorbent assay (ELISA)

2.12.1 VEGF and soluble endoglin

Serum levels of mouse VEGF and soluble endoglin (sEng) were determined with the VEGF DuoSet® ELISA (R&D Systems, DY493) and mouse Endoglin DuoSet® ELISA (R&D Systems, DY1320). All incubations were performed at RT and the plate was washed after each incubation step with 0.05% Tween-20 in PBS. After coating a microtiter plate (Nunc-Immuno™ Plate, 456537) with capture antibody overnight, the plate was blocked with 1% BSA in PBS. A 2-fold dilution series (1000 to 15.625 pg/ml VEGF; 4000 to 62.5 pg/ml sEng) was prepared for generating a standard curve and mouse serum samples were diluted 1:2 in 1% BSA and PBS. Standards and samples were incubated for 2 h. After the incubation for 2 h with detection antibody, Streptavidin-HRP was applied for 20 min. 3,3',5,5'-Tetramethylbenzidine (TMB) substrate solution incubated for 20 min in the dark and the colorimetric reaction was stopped by adding 4N H_2SO_4. The optical density (OD) was determined at 450 nm. VEGF and sEng serum levels were calculated from the respective standard curves with a four parameter logistic (4-PL) curve-fit.

2.12.2 Soluble fms-like tyrosine kinase-1

The soluble fms-like tyrosine kinase-1 (sFlt-1) serum levels were measured with the mouse soluble Flt-1 Quantikine Immunoassay (R&D Systems, MVR100). All incubations were performed at RT. The pre-coated microtiter plate was incubated for 2 h with the supplied Flt-1 standards for generating a seven point standard curve (125 to 8000 pg/ml) and mouse serum samples (1:2 dilution on gd 7.5; 1:4 on gd 15.5). After incubating the HRP-conjugated detection antibody for 2 h, the TMB substrate solution was added and incubated for 30 min in the dark. The colorimetric reaction was stopped with diluted HCl. The OD was determined at 450 nm and the sFlt-1 serum levels were calculated from the standard curve with a four parameter logistic (4-PL) curve-fit.

2.12.3 Galectin-1

Circulating human gal-1 levels were measured with a specific sandwich ELISA [65]. The microtiter plate (Nunc-Immuno™ Plate, 449824) was coated overnight with an anti-human gal-1 antibody (2 mg/ml; R&D Systems, AF1152) and washed after every incubation step with 0.5% Tween-20 in PBS. All incubations were performed at RT. After blocking the plate with 1% BSA in PBS for 2 h on a shaker, the wells were incubated with a 2-fold dilution series of recombinant human gal-1 (156.25 to 10,000 pg/ml; R&D Systems, 1152-GA) for generating a standard curve and with diluted serum samples for 2 h. Twenty-fold dilutions were prepared in blocking buffer. The biotinylated anti-human gal-1 antibody (0.25 µg/ml; R&D Systems, BAF-1152) was incubated for 2 h followed by a 30 min incubation step with HRP-conjugated streptavidin (Calbiochem, 189733). The TMB substrate solution was added for 20 min (dark) and the colorimetric reaction was stopped with 4N H_2SO_4. The OD was determined at 450 nm. Gal-1 serum levels were calculated from the standard curve with a four parameter logistic (4-PL) curve-fit.

2.13 Angiotensin II receptor type 1 autoantibodies

Autoantibodies against the angiotensin II receptor type 1 (AT1AA) were determined in mouse serum [144]. Ammonium sulphate precipitation was used for the isolation of the immunoglobulin (IgG) fraction. For 1 ml serum, 660 µl of saturated ammonium sulphate was added drop-wise, mixed and incubated over night at 4 °C. After centrifugation (7 min, 5,000 x g, 4 °C), the sediment was dissolved in 500 µl PBS. To precipitate the IgG fraction, 500 µl ammonium sulphate was added for 1 min. This procedure was repeated from the centrifugation step. After the last centrifugation, the IgG fraction was dissolved in 660 µl PBS and dialysed against PBS for 24 h. The isolated IgG fractions were prepared at a 1:40 dilution in duplicates and added to cultured neonatal rat heart muscle cells. The activation of AT1 influences the chronotropic response of these cells. This effect was measured by counting the beating rate for 1 min at six different points within in the cell culture flask. The differences between the beating rates before and after the addition of IgG fractions were calculated. To prove that an increase in beating was AT1 antibody-specific, 1 µM of the AT1 antagonist Losartan was added and the beating rate was re-analysed. As an additional control, the IgG fraction was mixed with a synthetic peptide (kindly provided by Dr Gerd Wallukat, MDC Berlin, Germany) corresponding to the sequence of the second extracellular loop of the AT1 (AFHYESQ) before measuring the chronotropic response. If no increase in the beating rate was observed for a serum sample, 1 µM angiotensin II, which activates the AT1, was added as a functional control of the experiment.

2.14 Kidney function assessment

2.14.1 Evaluation of kidney filtration capacity

For the evaluation of the filtration capacity, mice received an i.v. injection of 2.5 mg FITC-labelled dextran (Sigma-Aldrich, FD-2000S) 15 min before euthanasia [145]. Kidneys were frozen for cryo sectioning at 10 µm. Sections were counter-stained with DAPI and the renal glomeruli were examined with the fluorescence microscope. In a healthy kidney, the FITC-labelled dextran is filtered in the glomeruli – comparable to proteins – from the blood due to its high molecular weight (MW 2,000,000), so that a green staining is detected. In a kidney with a reduced filtration capacity, which can be caused by endotheliosis, less FITC-labelled dextran is filtered and is eliminated in the urine instead.

2.14.2 Proteinuria

For the determination of potential proteinuria, the albumin concentrations were measured with the Albuwell M ELISA (Exocell, #1011) in 24 h mouse urine samples. The urine was cleared by centrifugation and the murine albumin standards were prepared in a 2-fold dilution series (from 10 to 0.156 µg/ml). The standards and samples were incubated on the supplied test microtiter plates with the anti-albumin antibody for 30 min at RT. The HRP-conjugated antibody incubated for 30 min. The plates were washed after each incubation step. After 10 min colour development, the OD was determined at 450 nm and albumin concentrations of the urine samples were calculated from the standard curve.

2.15 Blood pressure measurements

The blood pressure was measured in the tail artery of pregnant female mice with a computerised, non-invasive tail-cuff acquisition system (CODA System, Kent Scientific Corporation)[146]. Before starting the actual measurements, the mice were trained in the blood pressure system for 30 min on three days to reduce stress. The body temperature of the mice was maintained between 34 and 36 °C with infrared heating while they were placed in the plastic holders and they had 15 min for acclimatisation before starting the data acquisition. An adequate blood flow in the tail is important for correct measurements. The system detects changes in the blood volume in the tail by Volume Pressure Recording (VPR) and measures systolic, diastolic and mean blood pressures, heart pulse rate, tail blood volume and flow. Ten measurement cycles were performed for each mouse and experimental day. Measurements were validated by optical control of the amplitudes.

2.16 *In vitro* assays

All cells used in the *in vitro* assays were a kind gift from Dr Judith E. Cartwright (St. George's University of London, UK). **SGHPL-4 cells** (St George's Hospital Placental cell Line-4) are derived from primary human first trimester extravillous trophoblasts and transfected with the early region of SV40, previously known as MC418. These cells resemble extravillous trophoblasts in their morphological and functional properties [147]. Tube formation was assessed in growth factor-reduced Matrigel in µ-Slides (10 µl per well). SGHPL-4 cells were maintained in serum-free media and were seeded onto Matrigel-coated wells (10,000 cells per well). Cells were untreated (control) or treated with 10 and 20 µM anginex, respectively, and incubated for 8 h at 37 °C. **Tube formation** and **branching points** were assessed through an inverted phase-contrast microscope at 5x magnification and the quantification was done with the WimTube Software (Wimasis, Germany). For the assessment of SGHPL-4 cell adherent capacity, cells were stained with CMFDA and incubated with 10 µM anginex for 24 h. Trypsinised cells (2.5×10^4) were seeded on confluent SGHEC-7 cells (St George's Hospital Endothelial Cell Line-7; 5×10^4 cells) in 96-well black/clear plates. Non-adherent cells were removed after 1 h and **adhesion** was analysed with the microscope and ImageJ software (NIH, USA). Density histograms were generated to calculate pixels. **Apoptosis** was determined with the Caspase-Glo® 3/7 Assay (Promega, G8090). The SGHPL-4 cells were cultured in 96-well plates with 10 µM anginex for 2, 5, 7, or 24 h. Staurosporine (10 nM) activates caspase-3 thereby inducing cell apoptosis and served as a positive control in this experiment. The luminescent signal was detected within 1 h after cell lysis.

2.17 Microarray

Microarrays provide the possibility to analyse many thousands of transcripts within one experiment and to compare the expression levels between experimental groups. Untreated and anginex-treated (5 and 24 h) SGHPL-4 cells (described in section 2.16) were analysed with the Illumina® HumanHT-12 v3 Expression BeadChip (BD-103-0203) according the Minimum Information About a Microarray Experiment (MIAME) criteria [148].

Total RNA was isolated as described in section 2.11, transcribed and labelled using the Illumina® TotalPrep™ RNA Amplification Kit (IL1791) in biological quadruplicates for all samples. Briefly, 50 ng total RNA in 11 µl Nuclease-free Water were incubated for 2 h at 42 °C with the Reverse Transcription Master Mix (first strand cDNA synthesis). Second Strand Master Mix was added and incubated for 2 h at 16 °C. The cDNA was purified with 250 µl cDNA Binding Buffer in a cDNA Filter Cartridge followed by 500 µl Wash Buffer and elution in 20 µl pre-heated (55 °C) Nuclease-free Water. The cDNA was transcribed to cRNA with 7.5 µl IVT Master Mix for 4 to 14 h at 37 °C. After the addition of 75 µl Nuclease-free Water, samples were mixed with 350 µl cRNA Binding Buffer and 250 µl 100% EtOH and passed through cRNA Filter Cartridges. The cRNA was washed with 650 µl Wash Buffer and eluted with 200 µl pre-heated Nuclease-free Water.

The cRNA was hybridized to the HumanHT-12 v3 Expression BeadChip array and gene expression was measured in all samples. The on probe levels were normalised without background correction with the Illumina® GenomeStudio® software (v2011.1 with Gene Expression Module v1.9.0). Data were log2 transformed after an offset addition and probes failing to underrun a detection p-value of 0.05 were discarded. An outlier check was performed for all samples using principal component analysis (PCA). Probes were analysed for significant expression differences in time course with ANOVA followed by FDR multiple testing correction [149]. Probes undergoing 5% FDR and exceeding a 1.5-fold change in time course were selected as differentially expressed (785 probes) and were K-Mean clustered using an Euclidean distance function and K = 5.

2.18 Statistical analyses

Statistical analysis was performed with Prism Software (version 5.0, GraphPad Software, Inc., USA). Data are expressed as mean ± SEM. Mouse and human data were analysed by nonparametric Mann-Whitney U-test and ANOVA. Differences among groups were evaluated using the Kruskal-Wallis, Tukey's or Bonferroni test. A p-value less than 0.05 was considered as statistically significant.

3. Results

3.1 Gal-1 acts as a pro-angiogenic factor during early gestation in mice

In this study, a CD11c.DTR mouse model was used to define the role of gal-1 during angiogenic processes associated with early pregnancy. CD11c.DTR female mice display a reduced vascular development after the ablation of dendritic cells (DC) with Diphtheria toxin (DT) [82]. We supplemented gal-1 exogenously in these mice (Figure 6). To demonstrate that gal-1 signalling is mediated via the VEGFR2 pathway, this receptor was blocked with a neutralizing antibody (clone DC101).

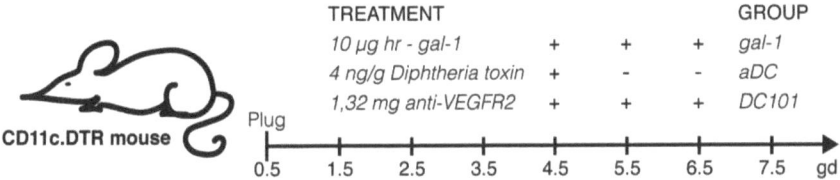

Figure 6: Experimental design in a model of reduced vascular development and attenuated expansion and maturation (CD11c.DTR mouse model)
CD11c.DTR mice were treated with 10 μg human recombinant galectin-1 (hr-gal-1) / injection on the respective gestation day (gd) to exogenously supplement gal-1 levels. For the ablation of dendritic cells (aDC) over the peri-implantation period, pregnant females received injections of 4 ng/g body weight Diphteria toxin (DT). The signalling of gal-1 via the VEGFR2 pathway was blocked with the injection of 1.32 mg VEGFR2 antibody (clone DC101). Mice were treated with the substances alone or in different combinations (cf. Table 2 in Methods).

First, we analysed the histology of gd 7.5 implantation sites by H&E staining. A similar histological appearance was detected in control females injected with either sterile PBS or gal-1, showing that the administration of gal-1 alone had no effect in CD11c.DTR mice (Figure 7a). In mice with ablated DC (aDC), the implantations were smaller, disorganised, contained no viable embryonic cavity and were considered as implantation failure. Gal-1 supplementation in aDC females recovered the normal histology and thus viability of the implantation site. This effect was abrogated when gal-1 signalling was blocked via the VEGFR2 pathway with a neutralizing antibody (clone DC101) (Figure 7b).

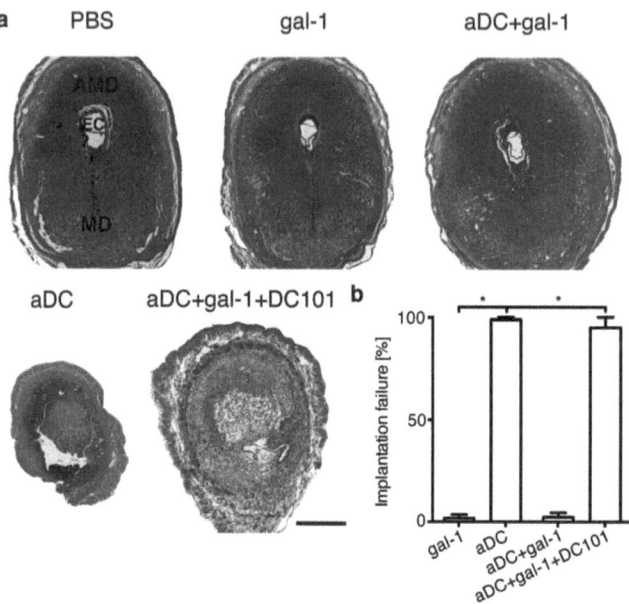

Figure 7: Gal-1 prevents implantation failure in mice with ablated DC
a) H&E staining allowed the histological analysis of gd 7.5 implantation sites (n = 5-6). Healthy implantations display organised mesometrial and anti-mesometrial deciduas (MD and AMD) and a well-formed embryonic cavity (EC). Scale bar = 500 µm. b) Implantation failure was evaluated macroscopically when dissecting gd 7.5 uterine tissue and calculated as follows: implantation failure = (non-viable implantations x 100) / total number of implantations (n = 6-8). *, P < 0.05.

To investigate the contribution of gal-1 to early angiogenic processes during pregnancy, we used the endothelial cell markers CD31 (PECAM-1) and endoglin to label gd 7.5 implantations with immunofluorescence. CD31 is a marker for existing mature blood vessels and was expressed in the part of the vascular zone that is close to the embryo, where angiogenic processes are completed on gd 7.5 (Figure 8a). This pattern was disturbed by DC ablation, but could be restored with gal-1 supplementation.

Endoglin is expressed in blood vessels newly created by angiogenesis in the vascular zone (Figure 8b). aDC mice were characterised by a disorganised vascular zone, seen by endothelial cells scattered throughout the whole implantation and a reduced staining intensity. Gal-1 rescued the endoglin expression pattern in these mice.

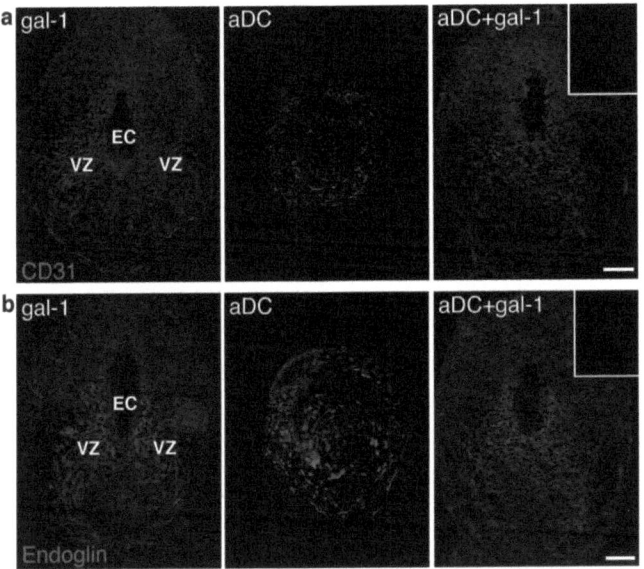

Figure 8: Ablation of DC during early pregnancy causes a reduced vascular development, which is prevented by gal-1 supplementation
a) CD31 is used as marker for mature blood vessels and was analysed by immunofluorescence on gd 7.5 (n = 5). Normally, CD31 is expressed in the vascular zone that is close to embryo. Inset shows negative control. Scale bar = 250 µm. b) Endoglin is expressed in angiogenic active blood vessels in the vascular zone of mouse implantations on gd 7.5 (n = 5). Scale bar = 250 µm. EC: embryonic cavity, VZ: vascular zone.

For a closer investigation of angiogenic status, we used an angiogenesis array to assess 53 angiogenesis-related proteins in aDC and aDC+gal-1 implantations. Gal-1 supplementation up-regulated 15 proteins that are mainly involved in matrix remodelling and other processes promoting angiogenesis (Table 8, Figure 9a). Additionally, aDC mice tended towards a reduction in serum VEGF levels when compared to gal-1 and aDC+gal-1 females on gd 7.5 (Figure 9b). In line with this, gal-1 increased the bioavailability of VEGF by reducing serum levels of the anti-angiogenic factor sFlt-1 on gd 7.5 (Figure 9c).

Table 8: Gal-1 up-regulated proteins involved in the angiogenesis process during early pregnancy
The expression of 53 angiogenesis-related proteins was tested in tissue lysates extracted from whole implantations on gd 7.5. The lysates of 3 implantations were pooled per group. Fifteen proteins were increased in aDC mice upon gal-1 supplementation.

Protein symbol	Full name	Function
MMP-9	matrix metallopeptidase 9	involved in angiogenesis together with FGF basic and TIMP-1 [150,151]
PAI-1	plasminogen activator inhibitor-1; Serpine1	promotes angiogenesis [152]
PTX-3	pentraxin related gene	involved in angiogenesis together with FGF basic [153]
FGF basic	fibroblast growth factor basic; Fgf2	involved in angiogenesis together with TIMP-1 and MMP-9 [150,151]
IL-1α	interleukin 1 alpha	induces angiogenesis [154]
MMP-3	matrix metallopeptidase 3	regulation of matrix degradation and angiogenesis [155]
MMP-8	matrix metallopeptidase 8	regulation of matrix degradation and angiogenesis [151]
TIMP-1	tissue inhibitor of metalloproteinase 1	involved in angiogenesis together with FGF basic and MMP-9 [150,151]
ANG	angiogenin	induces neovascularisation [156]
ADAMTS1	a disintegrin-like and metallopeptidase (reprolysin type) with thrombospondin type 1 motif, 1	involved in matrix remodelling [157]
PLF	proliferin	stimulates angiogenesis [14]
IGFBP9	Insulin-like growth factor binding protein-9; NOV, CCN3	induces angiogenesis [158]
HB-EGF	heparin-binding EGF-like growth factor	induces angiogenesis by activating MMP-3 and -9 [159]
CXCL16	chemokine (C-X-C motif) ligand 16	involved in chemotaxis of endothelial cells [160]
LEP	leptin	stimulates endothelial cell migration and VEGF expression [161]

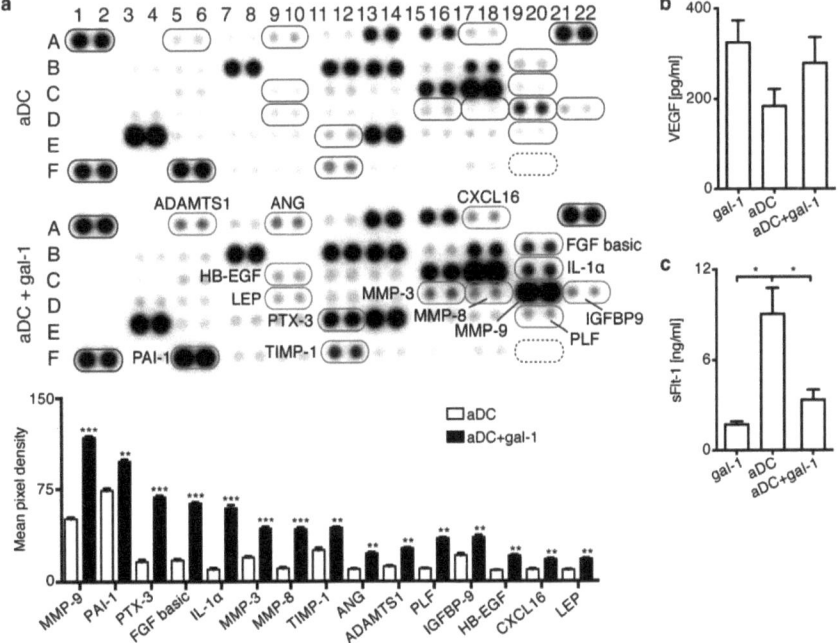

Figure 9: Gal-1 boosts an angiogenic milieu in aDC mice
a) Membranes were labelled with 53 different capture antibodies (in duplicates) and incubated with isolated proteins from gd 7.5 implantations of aDC (blue circles) and aDC+gal-1 (red circles) mice. The lysates of 3 implantations were pooled per group. Each spot represents an angiogenesis-related protein. Reference spots are encircled in black, negative controls with dashed black lines. The quantification showed that gal-1 administration in aDC mice up-regulated the protein expression of 15 genes (for details see Table 8). b) Maternal serum levels of VEGF were analysed by ELISA on gd 7.5 (n = 3-4). c) The sFlt-1 serum levels were determined with ELISA on gd 7.5 (n = 3-4). An increase in sFlt-1 levels corresponds to a reduction in the bioavailability of VEGF. *, P < 0.05; **, P < 0.01; ***, P < 0.001.

Since gal-1 was able to rescue aDC implantations, we next focussed our analysis on gd 10.5 placental development in both groups. In the placenta, foetal and maternal vessels come into close contact to exchange oxygen and nutrients to supply the embryo. The extracellular matrix of foetal vessels can be visualised by Isolectin B4 reactivity (Figure 10a) [162]. The placental depth was not changed in relation to the decidual size (Figure 10b). The area in the labyrinth that was occupied by foetal vessels was increased in aDC+gal-1 placentas (Figure 10c). However, the number of branches of foetal and maternal blood vessels did not differ between the groups (Figure 10d).

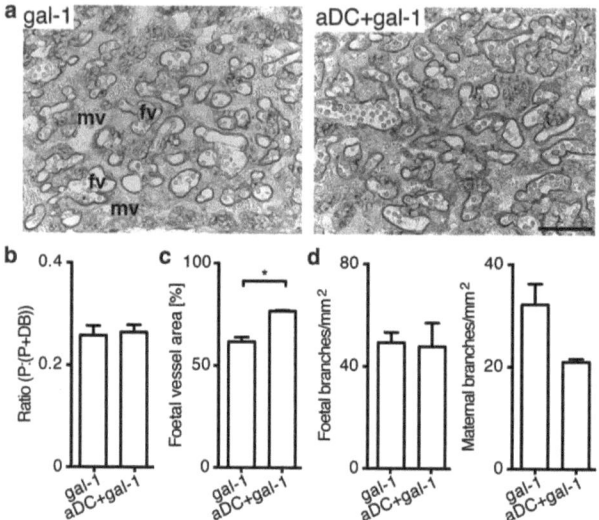

Figure 10: Gal-1 supports vascular development in the placenta of mice with ablated DC
a) Foetal vessels were visualised with Isolectin B4 (IB4) on gd 10.5. Scale bar = 100 μm. mv: maternal vessel, fv: foetal vessel. b) The placental and decidual sizes were measured and the ratio (P:(P+DB)) was calculated to analyse the relative placental depth (n = 5). c) The IB4$^+$ area occupied in the labyrinth zone of the placenta was analysed with ImageJ software (n = 5). d) The branches of foetal and maternal vessels in the labyrinth zone were counted in 3 squares (at 1 mm^2) per mouse (n = 5). *, P < 0.05.

It was additionally shown that the number of glycogen cells, which are located in the spongiotrophoblast, was not significantly reduced in aDC+gal-1 mice (Figure 11a). However, incomplete remodelling of the spiral arteries resulted in a higher wall thickness in aDC+gal-1 deciduas (Figure 11b).

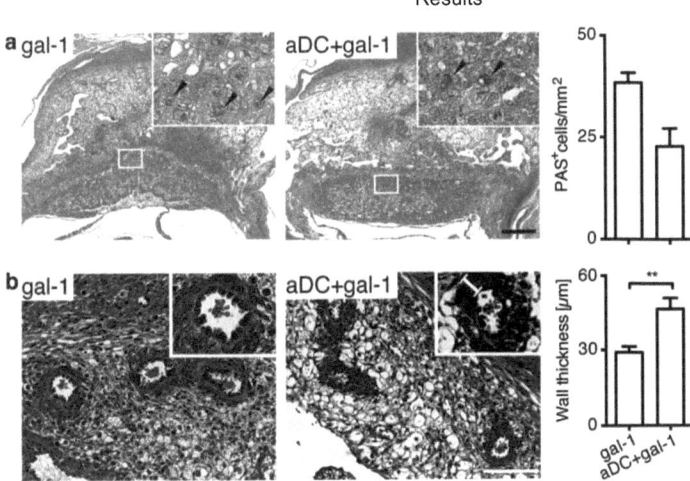

Figure 11: Gal-1 facilitates the placental development affected by DC ablation
a) Glycogen cells in the spongiotrophoblast of the placenta were stained with PAS. The number of glycogen cells was counted in 3 squares (at 1 mm^2) per mouse (n = 5). Arrows in insets denote PAS$^+$ glycogen cells. Scale bar = 500 μm. b) The spiral arteries were visualised with Masson-Goldner's trichrome stain and the wall thickness (denoted with ⊢⊣ in right inset) was determined (n = 5). At least 3 arteries were measured per mouse. Insets show single spiral arteries. Scale bar = 200 μm. **, $P < 0.01$.

Hormones such as placental lactogens modulate the maternal metabolism to support the foetal energy supply, therefore we analysed their expression in our mouse model. Placental lactogen-I (PL-I) is expressed in trophoblast giant cells (GC) during mid-gestation in control mice as previously described [163], while its expression is reduced in aDC+gal-1 mice (Figure 12a). Thereby, the total number of GC did not change. Another important placental hormone is proliferin (PLF), acting as a chemoattractant for endothelial cells of maternal vessels in the decidua [14]. aDC mice supplemented with gal-1 depicted a similar PLF expression within the GC layer compared to control mice (Figure 12b).

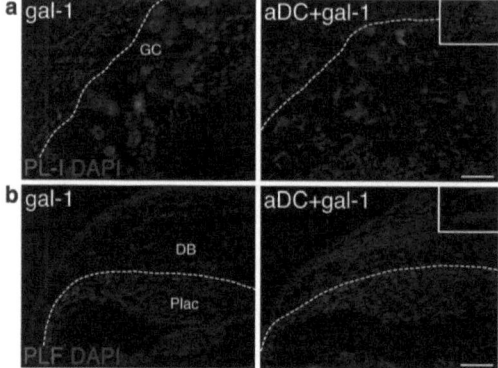

Figure 12: Placental lactogen-I and proliferin expression in giant cells upon gal-1 supplementation
a) Placental lactogen-I (PL-I) expression was analysed by immunofluorescence (n = 4-5). On gd 10.5, PL-I is expressed in trophoblast giant cells (GC). The white line marks the histological limit between the GC layer and decidua basalis. Inset represents the negative control. Scale bar = 100 µm. b) Proliferin (PLF) expression in the GC layer was determined by immunofluorescence on gd 10.5 (n = 4-5). The white line marks histological limit between the placenta (Plac; including the GC layer) and the decidua basalis (DB). Scale bar = 500 µm.

Next, we analysed maternal and foetal parameters of the mouse model regarding overall pregnancy outcomes. First, pregnancy fitness describes the number of females with a vaginal plug that become pregnant (implantation sites can be evidenced on gd 10.5, 13.5 and 16.5). Control females exhibited a pregnancy fitness between 90 and 100%. Although aDC+gal-1 mice showed a reduction in their pregnancy fitness on gd 13.5 and 16.5, supplementation with gal-1 supports the progress until late pregnancy in aDC mice (Figure 13a). The second maternal parameter describes the number of healthy implantations in a pregnant female, i.e. foetal loss or abortion rate, and was designated as pregnancy success. aDC+gal-1 females had a similar pregnancy success when compared to control dams (Figure 13b). Foetal development was assessed by body weight measurements and Theiler stage analysis on gd 16.5. Gal+1 and aDC+gal-1 embryos displayed a comparable body weight (Figure 13c). The developmental stage of these mice was evaluated with Theiler stage analysis. Mice at gd 16.5 are usually in Theiler stage 24 to 25. All analysed gal-1 and aDC+gal-1 embryos were within these developmental stages with fused eyelids, the pinna covering the external auditory meatus, parallel fingers and toes, a thickened skin with wrinkles, and a disappeared umbilical hernia (Figure 13d).

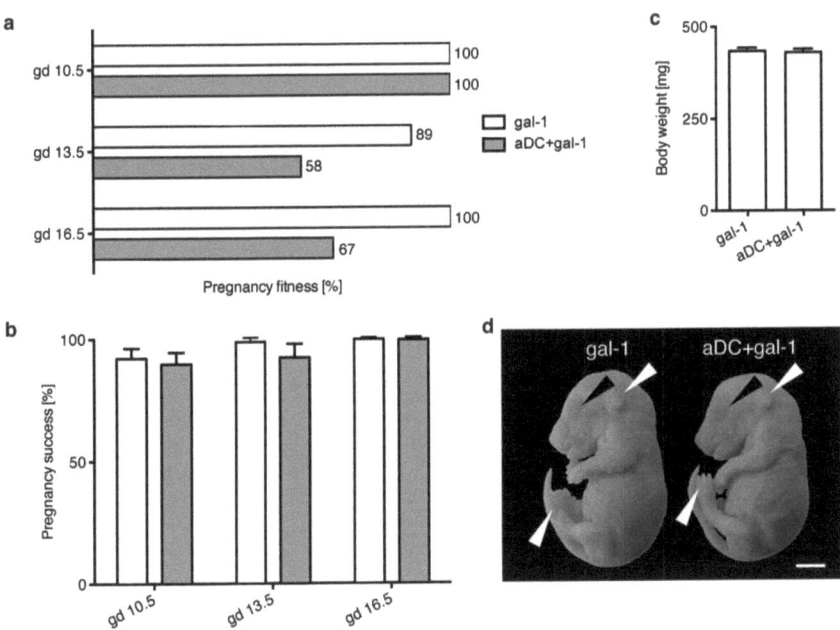

Figure 13: Gal-1 favours pregnancy progression and embryonic development in aDC mice
a) The pregnancy fitness describes how many females of a group and with a vaginal plug were actually pregnant at a certain gd and was calculated as follows: pregnancy fitness [%] = (pregnant females x 100) / all females with vaginal plug. We determined the pregnancy fitness on gd 10.5, 13.5, and 16.5 (n = 8-12 mated mice per group). b) The pregnancy success describes the number of healthy implantations per female and was calculated on gd 10.5, 13.5, and 16.5 as follows: pregnancy success = 100 – ((foetal resorptions x 100) / total number of implantations) (n = 6-8 per group). c) On gd 16.5, the body weights of gal-1 and aDC+gal-1 mice were analysed (n = 21-26 embryos). d) Theiler stage analysis was used to analyse the developmental stage of embryos from gal-1 and aDC+gal-1 mice on gd 16.5 (corresponding to Theiler stage 24 to 25), which is characterised by closed eye lids, the pinna covering the external auditory meatus and parallel toes (arrows) (n = 21-26 embryos). Scale bar = 0.25 cm.

Taken together, gal-1 was able to ameliorate the angiogenic state in a mouse model of reduced vascular expansion during the window of implantation. The early angiogenic signals were mediated via the VEGFR2 pathway and gal-1 thereby maintained a healthy gestation and embryonic development.

3.2 Inhibition of gal-1–mediated angiogenesis provokes PE-like symptoms

Knowing that gal-1 promotes the angiogenesis process, we next investigated to what extent the newly described pro-angiogenic function of gal-1 is critical for gestation. We thus inhibited the gal-1–mediated angiogenic function by daily injecting anginex from gd 4.5 to 12.5 (Figure 14). Anginex is a synthetic 33-mer cytokine-like designed peptide that binds to gal-1 and thereby blocks angiogenesis *in vivo* [97,102].

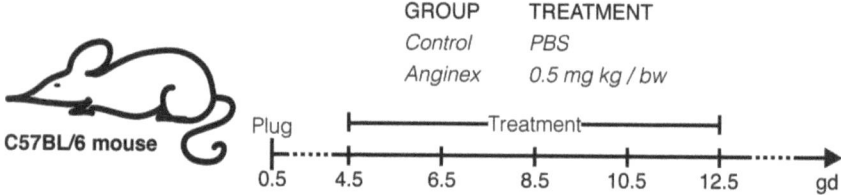

Figure 14: Experimental mouse model to inhibit the pro-angiogenic function of gal-1 with anginex
C57BL/6 pregnant females were treated with sterile PBS (control group) or 0.5 mg/kg body weight anginex from gd 4.5 to 12.5. Mice were killed on gd 13.5 or 15.5.

Anginex-treated mice exhibited similar foetal loss rates during gestation when compared to control dams (Figure 15a+b). However, embryos carried by anginex-treated females suffered from intrauterine growth restriction (IUGR) with reduced body weights and a delayed development corresponding to Theiler stage 23 instead of 24 (Figure 15c+d). The embryos of anginex-treated dams displayed open eyelids, a prominent umbilical hernia, no parallel fingers and toes, and the pinna not covering the external auditory meatus.

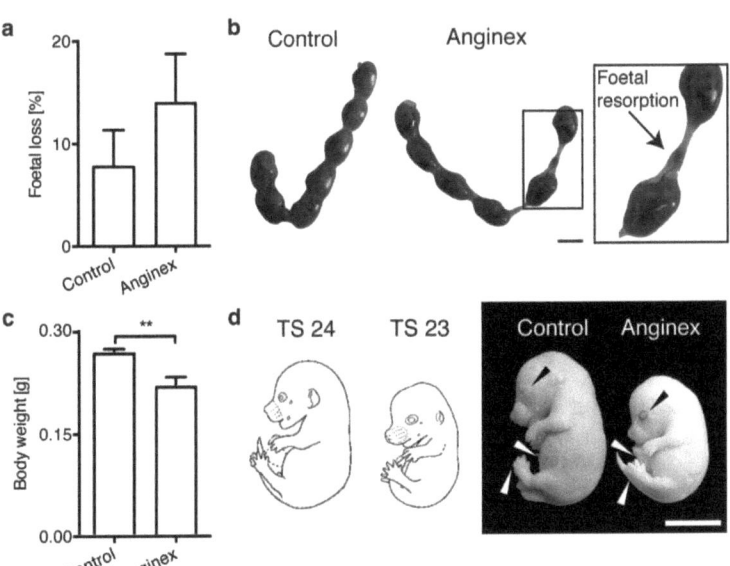

Figure 15: Mice treated with anginex suffer from IUGR
a) Foetal loss was analysed on gd 13.5 in anginex-treated and control dams (n = 6-8). Foetal loss = (foetal resorptions x 100) / total number of implantations. b) Pictures display complete implantation sites on gd 13.5 from one mouse at a time. A foetal resorption is magnified in the box. Scale bar = 1 cm. c) On gd 15.5, the embryo body weight was determined to identify intrauterine growth restriction (IUGR) (n = 9-20 embryos). d) To assess the foetal development, a Theiler stage (TS) analysis was performed in embryos on gd 15.5 (n = 9-20 embryos). Normal embryos on gd 15.5 are in TS 24. IUGR embryos are in TS 23 and are characterised by open eyelids, a prominent umbilical hernia, and no parallel fingers and toes (arrows). Scale bar = 0.5 cm. **, P < 0.01.

As mentioned previously, the placenta plays a main role during foetal development. In particular, maternal blood flow in the placenta is important to meet the foetus' nutrient requirements. Anginex-treated females displayed smaller placentas (Figure 16a). In this context, uNK cells are critical for decidual integrity and maternal spiral artery remodelling [81]. Mature uNK cells are defined by their DBA and PAS reactivity and are localised to blood vessels or found between decidual cells. Thus, we next focussed our analysis to the distribution and frequency of uNK cells on gd 13.5. In anginex-treated mice, the number of vascular uNK cells remained unchanged, while the tissue uNK cells were reduced (Figure 16b). Furthermore, the number of glycogen cells in the spongiotrophoblast of anginex-treated mice was also decreased (Figure 16c).

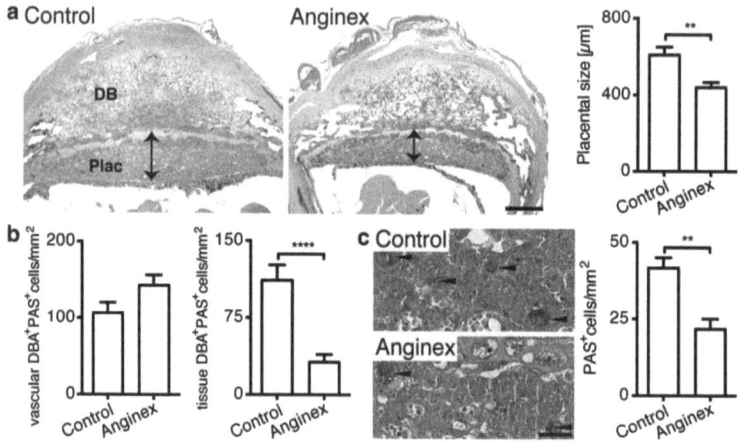

Figure 16: Inhibiting gal-1 pro-angiogenic functions compromises placental development
a) Isolectin B4 (IB4) staining was used to analyse the placental size (indicated by arrows) in implantations from control and anginex-treated dams on gd 13.5 (n = 5). Scale bar = 500 µm. DB: decidua basalis, Plac: placenta. b) The number of vascular and tissue uNK cells in the decidua basalis was analysed by DBA-PAS double staining on gd 13.5 (n = 5). Cells were counted in 3 to 4 squares (at 1 mm^2) per mouse. c) Glycogen cells (arrows) in the spongiotrophoblast were stained with PAS and counted in 3 squares (at 1 mm^2) per mouse on gd 13.5 (n = 5). Scale bar = 50 µm. **, P < 0.01; ****, P < 0.0001.

Next, hormones such as PL-I and PLF were analysed by immunofluorescence on gd 13.5. PL-I and PLF expression was reduced in the GC layer obtained from anginex-treated dams when compared to control mice (Figure 17a+b).

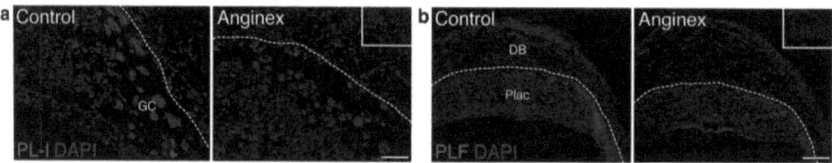

Figure 17: Placental hormones are reduced by anginex treatment

Results

a) Placental lactogen-I (PL-I) expression in giant cells (GC) was assessed by immunofluorescence staining on gd 13.5 (n = 5). The white line marks the histological limit between the giant cell (GC) layer and the decidua basalis. Inset represents the negative control. Scale bar = 100 µm. b) The expression of proliferin (PLF) in the GC layer was analysed by immunofluorescence on gd 13.5 (n = 5). The white line marks the histological limit between the placenta (Plac; including the GC layer) and the decidua basalis (DB). Scale bar = 500 µm.

During placentation the maternal spiral arteries in the decidua basalis are remodelled into large vessels to supply the growing demands of the embryo for oxygen and nutrients. This process is reduced in women suffering from PE. As shown in Figure 18a, analysis of Masson-Goldner's trichrome staining revealed that arterial walls remained thick after anginex treatment suggesting an impaired remodelling of the maternal spiral arteries. This could result in a low placental perfusion thereby causing hypoxia within the placenta, which is a typical feature during PE pathogenesis [118]. Therefore, we next analysed the oxygen tension *in vivo* with pimonidazole-HCl, which binds adducts that form under hypoxic conditions. On gd 13.5, anginex-treated mice displayed a lower oxygen tension, which is reflected by a more hypoxic GC layer detected in the placentas (Figure 18b).

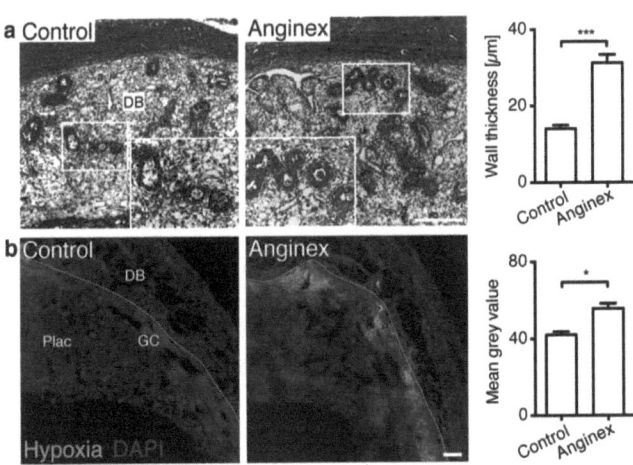

Figure 18: Blocking gal-1 pro-angiogenic function influences the oxygen tension in placental tissues
a) Masson-Goldner's trichrome stain of implantations on gd 13.5 shows the spiral arteries (insets) in the decidua basalis (DB) of control and anginex-treated dams (n = 5). For illustration of wall

thickness see Figure 11b. Scale bar = 250 μm. b) The oxygen tension in the placenta was determined with pimonidazole-HCl in vivo on gd 13.5 (n = 5). The white line marks the histological limit between the placenta (Plac; including the giant cell layer (GC)) and the decidua basalis (DB). The mean grey value was analysed with ImageJ software. Scale bar = 100 μm. *, P < 0.05; ***, P < 0.001.

We next focussed on two anti-angiogenic factors, sEng and sFlt-1, that are released together with placental cellular debris into the maternal circulation. These factors cause a systemic inflammatory response and dysfunctional endothelia during the course of PE and contribute to the clinical symptoms of PE during late pregnancy[114]. In line with this, maternal sEng levels in the circulation were elevated on gd 15.5 upon anginex treatment (Figure 19a). No changes were observed in the maternal sFlt-1 levels between control and anginex-treated dams. Knowing that sEng and sFlt-1 contribute to the manifestation of hypertension during PE [119,120], we measured the blood pressure in control and anginex-treated dams. Females that received anginex over gestation developed hypertension in late pregnancy (Figure 19b).

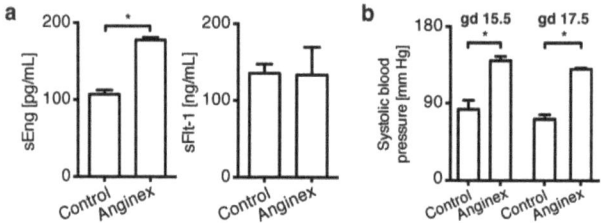

Figure 19: Gal-1 inhibition is correlated with higher circulating sEng levels and hypertension in late gestation
a) The levels of anti-angiogenic sEng and sFlt-1 were analysed in maternal serum by ELISA on gd 15.5 (n = 4). b) The systolic blood pressure was measured in the tail artery of dams on gd 15.5 and 17.5 to identify hypertension (n = 4-5). *, P < 0.05.

Patients that develop preeclampsia during pregnancy usually suffer from renal dysfunctions characterised by proteinuria. The H&E kidney staining in Figure 20a shows the renal histology and its glomeruli that are located in the renal cortex. Each glomerulus contains a network of capillaries, through which the blood is filtered. Figure 20b illustrates the renal function in control and anginex-treated dams as analysed by FITC-dextran filtration capacity on gd 15.5. We observed that females treated with anginex showed a diminished filtration capacity of FITC-dextran, evidenced by reduced green fluorescence in the glomeruli compared to control mice. In concordance with that, the anginex-treated females displayed proteinuria as analysed by the albumin concentration in collected 24 h urine between gd 14.5 and 15.5 (Figure 20c).

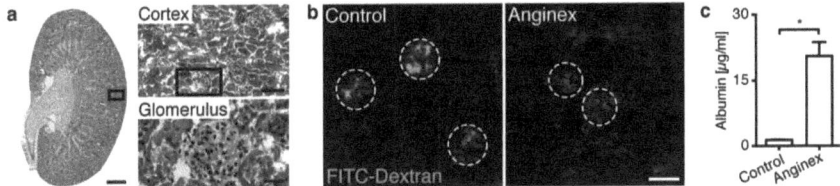

Figure 20: Renal dysfunction during late gestation is observed upon anginex treatment
a) H&E staining shows kidney histology with magnifications of renal cortex and glomerulus. Scale bars: 1000 µm, 200 µm (Cortex) and 50 µm (Glomerulus). b) The filtration capacity of the glomeruli (white circles) was analysed by injection of FITC-labelled dextran in vivo (n = 5). The green fluorescence in the glomeruli was analysed in kidney sections on gd 15.5 and corresponds to filtration capacity. Scale bar = 100 µm. c) The albumin concentration in 24 h urine from gd 14.5 to 15.5 was determined in control and anginex-treated dams by ELISA to detect potential proteinuria (n = 4). *, P < 0.05.

In order to better characterise the anginex-treated dams, we next analysed the renin-angiotensin system (RAS), which is a hormone system that regulates water balance and blood pressure. Angiotensin II receptor type 1 autoantibodies (AT1AA) are increased in PE patients contributing to the dysregulation of the RAS. This, in turn, leads to high oxidative stress and an excessive maternal inflammatory response that characterise the PE syndrome [144,164-166]. AT1AA levels in the gd 15.5 maternal serum immunoglobulin (IgG) fraction were determined by addition to cultured neonatal rat heart muscle cells. Analysis was performed by measuring the change in beating rate due to the AT1 activation in these cells. The serum of the control mice contained low levels of AT1AA; the increase over the basal level was below one beat per minute (Figure 21a). We added angiotensin II

(AngII) as a positive functional control and, as shown in Figure 21a, the beating rate increased due to the activation of the AT1 by AngII. In particular, serum from anginex-treated dams contained more AT1AA as evidenced by an elevated beating rate. This effect was fully blocked by the AT1 antagonist Losartan and partly blocked by a synthetic peptide that corresponds to the sequence of the second extracellular loop of the AT1 (AFHYESQ) and thus competes with the AT1 for binding the AT1AA. When analysing placental tissue, we observed that local RAS was also affected by anginex treatment showing a higher renin mRNA expression on gd 15.5 (Figure 21b).

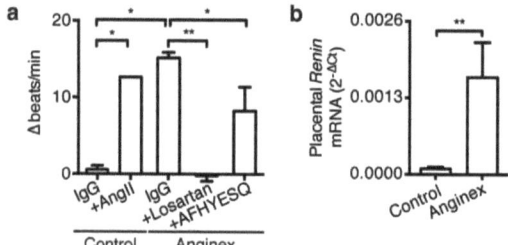

Figure 21: Anginex-treated mice show dysregulated RAS pathways
a) The maternal serum immunoglobulin (IgG) fraction was isolated by ammonium sulphate precipitation from control and anginex-treated dams on gd 15.5 (n = 4). To investigate if the serum contained angiotensin II receptor type 1 autoantibodies (AT1AA), the IgG fraction was added to cultured neonatal rat heart muscle cells, and changes in the beating rate due to AT1 activation were determined. Angiotensin II (AngII) served as a functional positive control. The observed change in beating rate was verified with the AT1 antagonist Losartan and the synthetic peptide AFHYESQ that competes with AT1 for AT1AA binding. b) The placental renin mRNA levels were determined by qPCR in control and anginex-treated dams on gd 15.5 (n = 5). *, P < 0.05; **, P < 0.01.

In summary, these results showed that the inhibition of gal-1–mediated angiogenic processes compromised the vascularisation and placentation during gestation and provoked PE-like symptoms in mice.

3.3 Gal-1 inhibition impairs human trophoblast functions *in vitro*

To further investigate if gal-1 inhibition and its effect on the placenta could contribute to the development of PE, human trophoblast SGHPL-4 cells, which resemble extravillous trophoblasts in their morphological and functional properties [147], were treated with anginex. Anginex reduced tube formation and number of branching points formed by SGHPL-4 cells in a dose-dependent manner, indicating diminished blood vessel formation (Figure 22a). The adherent capacity of trophoblast cells, which is important for the remodelling of spiral arteries, was evaluated by incubating anginex-treated SGHPL-4 cells with SGHEC-7 endothelial cells. As shown in Figure 22b, untreated SGHPL-4 cells have the capacity to adhere to the endothelial cell layer. In contrast, anginex-treated SGHPL-4 cells displayed a decreased adhesion capacity *in vitro*. In addition, we observed that anginex did not induce apoptosis of SGHPL-4 cells (Figure 22c), implying that the negative effects exhibited by anginex only influenced the functional properties, but not survival of trophoblast cells.

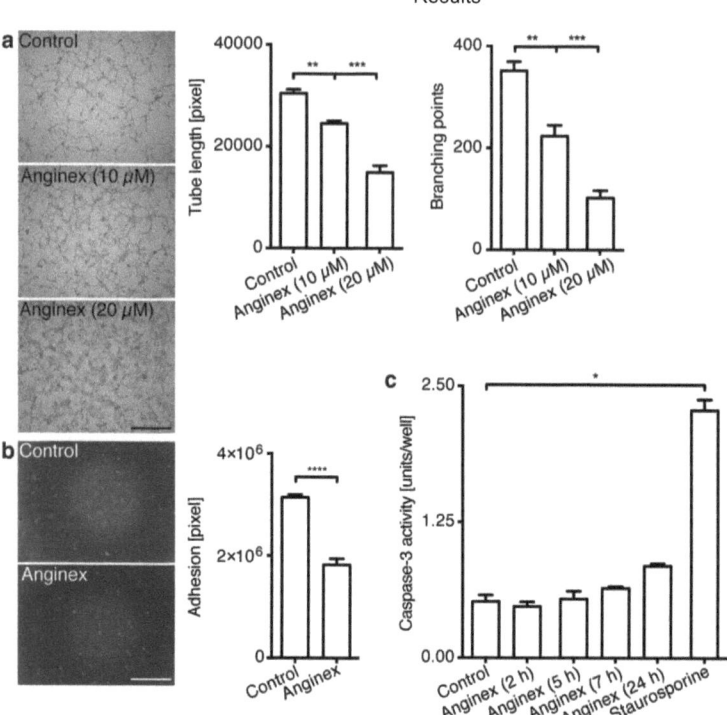

Figure 22: Trophoblast functions are affected by anginex treatment in vitro
a) Human trophoblast SGHPL-4 cells displaying extravillous trophoblast properties were seeded onto Matrigel-coated wells. Cells were untreated or treated with 10 and 20 µM anginex for 8 h (n = 4-9 per treatment), then tube formation and branching points were assessed with a microscope. Scale bar = 500 µm. b) CMFDA-stained SGHPL-4 cells were untreated or incubated with 10 µM anginex for 24 h and subsequently seeded on SGHEC-7 endothelial cells (n = 9 per treatment). The adhesion of SGHPL-4 cells (green) to SGHEC-7 cells was analysed with a microscope. Scale bar = 500 µm. c) SGHPL-4 cells were untreated or incubated with 10 µM anginex for the indicated times (n = 3-4 per treatment). The induction of apoptosis was analysed with a caspase assay. Staurosporine is an inducer of apoptosis and served as a positive control in this assay. *, $P < 0.05$; **, $P < 0.01$; ***, $P < 0.001$; ****, $P < 0.0001$.

Next, we performed a microarray analysis to identify dysregulated genes in anginex-treated SGHPL-4 cells (Figure 23). After 24 h treatment, 315 genes were up-regulated and 138 down-regulated. Thirty-eight genes were transiently up-regulated, i.e. they were increased after 5 h anginex treatment, but did not differ from the controls after 24 h. Finally, 67 genes displayed higher mRNA levels and 107 genes lower levels after 5 h anginex treatment. The expression of the up-regulated genes was found to be commonly induced by oxidative stress, inflammatory cytokines, and immune activation. The down-regulated genes tended to encode proteins specific for trophoblast function. A list of selected genes and their functions is given in the appendix (Table 12).

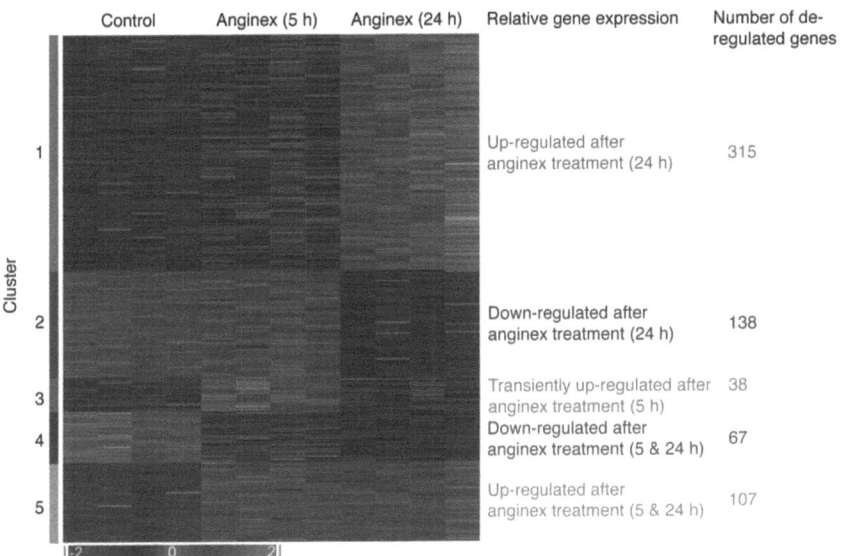

Figure 23: Anginex treatment provokes differential regulation of genes related to inflammation and angiogenesis in SGHPL-4 cells
The heat map displays the relative mRNA expression from SGHPL-4 cells that were untreated or treated with anginex for 5 or 24 h and analysed by microarray (n = 4 per treatment). Red represents a high and blue a low relative expression. Five clusters were defined depending on the relative gene expression.

In conclusion, the inhibition of gal-1 in human trophoblast cells impaired functional properties that are necessary for a normal placentation and adequate vascular development. Moreover, anginex dysregulated the expression of genes that are involved in biological processes related to PE.

3.4 *Lgals1* deficient mice spontaneously develop PE-like symptoms

To gain further insight into the role of gal-1 during PE, *Lgals1* wild type (wt) and deficient (knock out, ko) mice were investigated concerning the development of PE-like symptoms. The deficiency of gal-1 did not influence foetal loss during late gestation (Figure 24a). However, the embryos suffered from IUGR as shown by reduced body weight on gd 15.5 and 17.5 and delayed developmental stage at gd 15.5 (normally corresponding to Theiler stage (TS) 24) (Figure 24b+c). While the *Lgals1* wt embryos matched TS 24, the deficient mice resembled TS 23 as characterised by open eyelids, a prominent umbilical hernia, no parallel fingers and toes, and the pinna not covering the external auditory meatus. Additionally, the treatment with anginex had no effect on the body weight and Theiler stage of *Lgals1* deficient mice, demonstrating that anginex binds to gal-1 and thereby exerts its anti-angiogenic effect (Figure 24b+c).

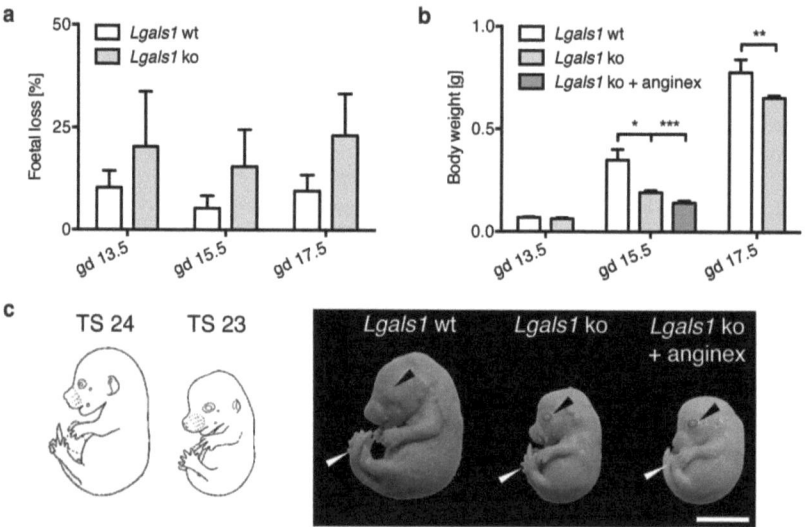

Figure 24: Lgals1 deficient embryos suffer from IUGR

Results

a) Foetal loss = (foetal resorptions x 100) / total number of implantations. Foetal loss was determined on gd 13.5, 15.5 and 17.5 (n = 6-8) in Lgals1 wt and deficient dams. b) The embryonic body weight was measured on gd 13.5, 15.5 and 17.5 in Lgals1 wt and deficient offspring to identify a potential IUGR (n = 11-21 embryos). c) To analyse foetal development, the Theiler stage (TS) was determined on gd 15.5 that normally corresponds to TS 24 (n = 11-12 embryos). Embryos suffering from IUGR resembled TS 23 with open eyelids and no parallel toes (arrows). Scale bar = 0.5 cm. Lgals1 wt: Lgals1 wild type, ko: knock out (deficient). *, $P < 0.05$; **, $P < 0.01$; ***, $P < 0.001$.

The expression of placental hormones is an important factor for supplying the embryo with nutrients. In *Lgals1* deficient placentas, the PL-I and PLF expression was reduced on gd 13.5 (Figure 25a+b).

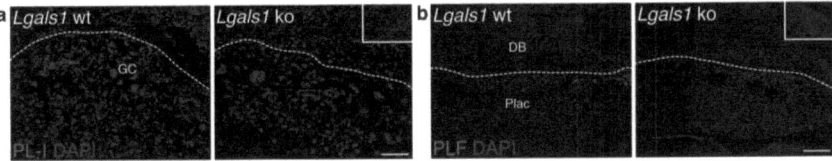

Figure 25: Placental hormone expression in Lgals1 deficient mice
a) The expression of placental lactogen-I (PL-I) in giant cells (GC) was assessed by immunofluorescence staining on gd 13.5 (n = 5). The white line marks the histological limit between the GC layer and the decidua basalis. Inset shows negative control. Scale bar = 100 µm. b) The placental expression of proliferin (PLF) in the GC layer was analysed by immunofluorescence on gd 13.5 (n = 5). The white line marks the histological limit between the placenta (Plac; including the GC layer) and the decidua basalis (DB). Scale bar = 500 µm.

To further test factors that influence the placental development, we assessed the expression of HIF-2α, placental growth factor (PGF), and VEGF by qPCR. HIF-2α is induced by low oxygen levels that occur during inflammatory processes and oxidative stress in the pathogenesis of PE [118,167]. HIF-2α mRNA levels were increased in *Lgals1* deficient placentas (Figure 26a). Moreover, the expression of PGF and VEGF, which is reduced in PE patients [125], was also decreased in *Lgals1* deficient dams. Further supporting this data, placental weights were also reduced in *Lgals1* deficient mice (Figure 26b).

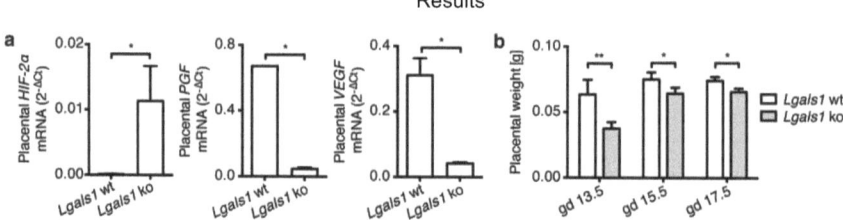

Figure 26: Placental development is retarded in Lgals1 deficient dams
a) On gd 15.5, placental expression of HIF-2α, placental growth factor (PGF), and VEGF mRNA was analysed by qPCR (n = 4-5). b) The placental weight of Lgals1 wt and deficient dams was determined on gd 13.5, 15.5 and 17.5 (n = 8-16 placentas). *, P < 0.05; **, P < 0.01.

During PE, angiogenic factors in the mother's circulation are dysregulated. In *Lgals1* deficient mice, pro-angiogenic VEGF tended to be lower but did not reach significance on gd 13.5 and 15.5 (Table 9). Additionally, the anti-angiogenic factors sFlt-1 and sEng, which reduce the bioavailability of VEGF, were elevated, although not significantly, in *Lgals1* deficient mice.

Table 9: Angiogenic factors in the circulation of Lgals1 mice
Circulating factors were determined in Lgals1 wt and deficient (ko) dams on gd 13.5 and 15.5 by ELISA (n = 3). Although no significant results were obtained, the data showed a tendency of reduced pro-angiogenic VEGF and increased anti-angiogenic sFlt-1 and sEng levels in Lgals1 deficient serum.

Gd & Group		VEGF	sFlt-1	sEng
gd 13.5	Lgals1 wt	120.94 ± 0.87 pg/ml	12.00 ± 3.86 ng/ml	849.16 ± 37.38 pg/ml
	Lgals1 ko	98.26 ± 16.92 pg/ml	20.80 ± 8.21 ng/ml	923.45 ± 55.18 pg/ml
gd 15.5	Lgals1 wt	125.00 ± 29.74 pg/ml	54.48 ± 22.44 ng/ml	1028.94 ± 199.55 pg/ml
	Lgals1 ko	100.37 ± 22.76 pg/ml	70.35 ± 18.67 ng/ml	1141.06 ± 200.35 pg/ml

Hypertension and proteinuria are the main symptoms of PE. Therefore, our next analysis comprised the evaluation of tail artery systolic blood pressure on gd 17.5 and an investigation of renal function. *Lgals1* deficient females displayed high blood pressure at the end of pregnancy (Figure 27a). Moreover, they suffered from proteinuria as shown by increased albumin levels in 24 h urine samples collected from gd 16.5 to 17.5 (Figure 27b).

Renal dysfunction was confirmed by FITC-dextran injection and its filtration through the glomeruli on gd 17.5. The filtration capacity was reduced in *Lgals1* deficient females (Figure 27c).

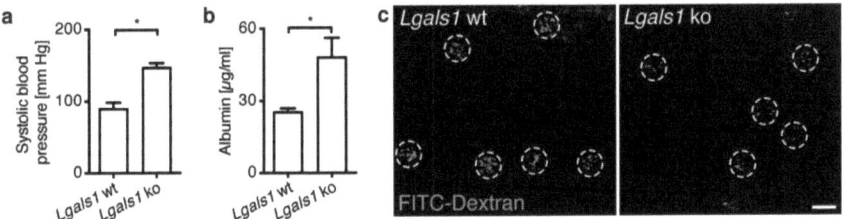

Figure 27: A PE-like syndrome characterised Lgals1 deficient dams
a) Systolic blood pressure was analysed by the tail-cuff method on gd 17.5 to identify hypertension (n = 4-5). b) To detect proteinuria, albumin levels were determined by ELISA in 24 h urine samples collected between gd 16.5 and 17.5 from Lgals1 wt and deficient dams (n = 4-5). c) The in vivo filtration capacity of the glomeruli (white circles) was determined by injection of FITC-labelled dextran on gd 17.5. The green fluorescence in the glomeruli was analysed in kidney sections (n = 5). Scale bar = 100 μm. *, P < 0.05.

Knowing that PE is also characterised by dysregulated systemic and placental RAS [144,168], we analysed the circulating levels of AT1AA and the placental expression of AT1 and renin in *Lgals1* wt and deficient dams during late pregnancy. The maternal serum IgG fraction containing AT1AA (gd 17.5) was added to cultured neonatal rat heart muscle cells and changes in the beating rate were observed. Serum from *Lgals1* deficient mice contained higher levels of AT1AA on gd 17.5 as evidenced by the increased change in the beating rate when compared to *Lgals1* wt dams (Figure 28a). The beating increase was prevented by the addition of the AT1 antagonist Losartan. In addition, the placental expression of AT1 and renin mRNA was up-regulated in *Lgals1* deficient mice on gd 15.5 (Figure 28b).

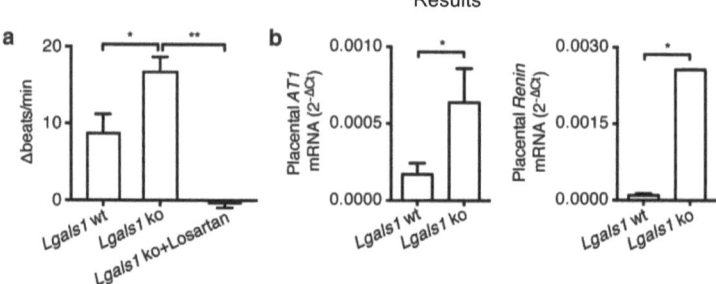

Figure 28: Systemic and placental RAS dysregulation is observed in Lgals1 deficient mice
a) Ammonium persulphate precipitation was used to isolate the immunoglobulin (IgG) fraction from the serum of Lgals1 wt and deficient dams on gd 17.5, which may contain angiotensin II receptor type 1 autoantibodies (AT1AA) (n = 4). After the addition of the IgG fraction to cultured neonatal rat heart muscle cells, the changes in the beating rate induced by AT1AA activation of AT1 were analysed. Since the AT1 antagonist Losartan could block the changes in the beating rate, this was considered to be AT1-dependent. b) On gd 15.5, the placental mRNA levels of renin and AT1 were analysed by qPCR to assess the regulation of placental RAS in Lgals1 wt and deficient dams (n = 4-5). *, $P < 0.05$; **, $P < 0.01$.

In summary, Lgals1 deficient mice spontaneously developed PE symptoms such as proteinuria, hypertension, dysregulated RAS, impaired placentation and IUGR during late pregnancy. Since the lack of gal-1 caused PE-like symptoms in both anginex-treated and Lgals1 deficient mice, it suggests that this lectin could be important during PE pathogenesis.

3.5 Placental gal-1 is dysregulated during PE in humans

The galectin mRNA profile in the utero-placental tissues was analysed by microarray in 23 healthy pregnant and 25 preeclamptic (PE) women in the group of Professor Dr Anne Cathrine Staff, Oslo University Hospital, Norway (Table 10). Only gal-1 was significantly down-regulated in PE patients. These results in combination with the previously described mouse data, concerning the role of gal-1 during the development of PE, prompted us to further investigate the gal-1 levels in humans. The women of the Oslo cohort did not undergo uterine contractions and had a caesarean section, so that no oxidative stress during delivery influenced gene expression. The cohort was divided into three groups according to their clinical characteristics: control women with uneventful pregnancies, preeclamptic women with early onset of PE and late PE (Table 3).

Table 10: Galectin profile in healthy pregnant women and PE patients
The expression of galectins in the utero-placental tissue from 23 healthy pregnant and 25 preeclamptic (PE) women was analysed by microarray in the group of Professor Dr Anne Cathrine Staff from Oslo University Hospital, Norway. The fold change was calculated; positive values denote an up-regulation, negative a down-regulation in control versus PE women. For the 14 galectin mRNAs tested, only gal-1 (LGALS1) was significantly reduced (P < 0.05, 1.3-fold) in PE patients.

Gene symbol	Chromosome	Illumina Gene ID	Fold change	p-value (t-test)
LGALS1	22	ILMN_1723978	-1.30618714	0.00314342
LGALS2	22	ILMN_1687306	-1.01674922	0.58618277
LGALS3	14	ILMN_1803788	-1.05920694	0.63280797
LGALS3BP	17	ILMN_1659688	-1.05806583	0.65458244
LGALS4	19	ILMN_1694034	1.06478834	0.53559651
LGALS7	19	ILMN_1661708	-1.02118454	0.50186293
LGALS8	1	ILMN_2353358	1.08984089	0.18801988
LGALS8	1	ILMN_1669930	-1.01900141	0.49789624
LGALS8	1	ILMN_2266214	-1.02818576	0.30053096
LGALS9	17	ILMN_1715760	-1.04413795	0.44969493
LGALS9	17	ILMN_2412214	-1.17815428	0.10606666
LGALS9B	17	ILMN_1656869	1.04207013	0.35408148
LGALS9B	17	ILMN_2113333	-1.01627945	0.56934962
LGALS9C	17	ILMN_2080342	1.00779302	0.80448568
LGALS12	11	ILMN_1776283	-1.02990046	0.55116589
LGALS13	19	ILMN_1794842	1.0499433	0.67244247
LGALS14	19	ILMN_2397223	-1.07449159	0.19925987
LGALS14	19	ILMN_1698318	-1.2125219	0.14639694

Next, we validated the microarray results by qPCR analysis of gal-1 expression in the human decidua and placenta of healthy pregnant women and PE patients. The decidual gal-1 levels were not changed between control, early and late PE groups (Figure 29a). In contrast, placental gal-1 mRNA was down-regulated in patients with early onset PE (Figure 29b). Furthermore, placental gal-1 protein expression was analysed by Western blot showing that gal-1 was reduced in early onset PE patients, but increased in late onset PE (Figure 29c). Immunofluorescence staining confirmed the differential gal-1 expression in the placental villous tissues in early and late onset PE patients, respectively (Figure 29d).

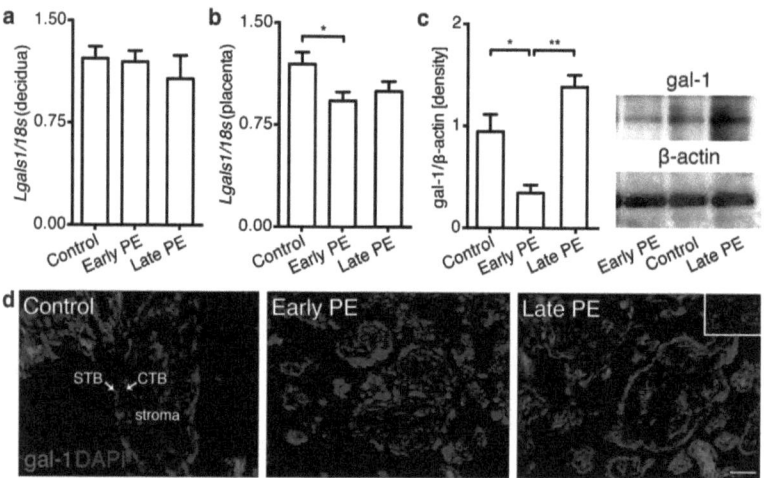

Figure 29: Differential placental gal-1 expression in early and late PE
a) Gal-1 levels were determined by qPCR in the deciduas of healthy pregnant and preeclamptic (PE) women at term pregnancy at Oslo University (Table 3; n = 17-36). PE women were separated into early and late disease onset groups (before or after 34th week of gestation, respectively). b) Placental gal-1 levels were analysed by qPCR in control, early PE and late PE pregnant women at term (n = 17-28). c) Gal-1 protein levels were determined by Western blot in placental tissues of healthy pregnant women and early and late onset PE patients (n = 5). d) Gal-1 protein expression in the placental villous tissue was also analysed by immunofluorescence (n = 5). Inset denotes negative control. Scale bar = 50 μm. STB: syncytiotrophoblasts, CTB: cytotrophoblasts. *, $P < 0.05$; **, $P < 0.01$.

Results

Circulating gal-1 was not changed in the early PE patients, but increased in late PE at the end of gestation (Figure 30a). We also tested blood samples from a prospective study at Brown University (Table 4, Figure 30b). The samples were taken at 22nd week of gestation (second trimester) from women with no clinical signs of PE, but some of them developed PE during later gestation. Interestingly, circulating gal-1 levels were reduced in the second trimester of these late PE patients and can thus predict the development of late PE.

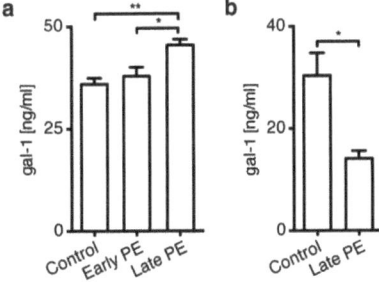

Figure 30: Circulating gal-1 levels predict the development of PE in human pilot studies
a) In the Oslo cohort, gal-1 levels were determined by serum ELISA in the third trimester of pregnancy (n = 9-15). b) The circulating levels of gal-1 were analysed in women during the second trimester in a cohort from Brown University, in which some women later developed late PE (Table 4; n = 7-8). *, $P < 0.05$; **, $P < 0.01$.

In humans, the local and peripheral gal-1 levels were dysregulated during the development of PE. Thereby, it is important to note that there were differences between the early and late onset PE. Human data confirmed mouse results, which demonstrated a main role of gal-1 in PE pathogenesis and its possible use as a predictor in second trimester blood screenings.

4. Discussion

Preeclampsia is one of the leading causes for maternal morbidity and mortality during pregnancy [111]. Although some progress has been made in the treatment of this syndrome, the only ultimate cure is the delivery of the baby, which bears the risk of complications for the foetus, depending on the gestational age. PE symptoms affect many maternal organ systems in different severities and the foetus usually suffers from IUGR [111,117]. Moreover, both the mother and the foetus carry a higher risk for the development of cardiovascular diseases during later life [169].

The underlying mechanisms that contribute to the pathogenesis of PE are not fully understood and diagnosis usually occurs late during pregnancy, when the symptoms have already manifested. For these reasons, there is the need for a biomarker allowing early detection of high-risk women. Once identified as high risk, women could undergo close screenings which would allow for early detection and treatment [111]. In this thesis, the angiogenic functions of gal-1 during pregnancy were closely investigated in different mouse models and its dysregulation before the clinical onset of PE was shown in a human pilot study.

The pro-angiogenic properties of gal-1 have been described in different physiological conditions [92,97]; however, the contribution of gal-1 to angiogenic processes during early gestation remained to be elucidated. In this thesis, we demonstrated, using a mouse model of reduced vascular expansion, that gal-1 increased the angiogenic status to maintain a successful pregnancy. Gal-1 was able to rescue the vascularisation in the decidua and promotes the development of the vascular zone. Both pre-existing and new blood vessels, as evidenced by the endothelial cell markers CD31 and endoglin, were supported by gal-1 supplementation. In addition, the angiogenesis array elucidated that gal-1 directly up-regulated proteins involved in matrix remodelling (e.g. MMPs) and endothelial cell migration, adhesion and proliferation (e.g. leptin, CXCL16, PLF, angiogenin). Gal-1 thus promoted angiogenic processes at the foetal-maternal interface during early gestation.

Moreover, gal-1 indirectly enhanced angiogenesis through the reduction of sFlt-1, which subsequently increased VEGF bioavailability. The VEGF receptors and NRP-1 are highly expressed in the mesometrial decidua, where the decidual angiogenesis and placentation take place [85]. Knowing that gal-1 promotes angiogenesis by binding to NRP-1, thereby activating VEGFR2 signalling, we blocked this pathway with the neutralizing antibody DC101 *in vivo* [86,88]. The DC101 antibody inhibits the ligand-induced VEGFR2 activation. The pregnancy-protective effect of gal-1 supplementation was abrogated and the mice suffered from complete implantation failure. Thus, we suggest that gal-1 mediates its pro-angiogenic functions during early gestation via NRP-1 and VEGFR2 signalling.

The rescue of decidual vascularisation by gal-1 facilitated a normal embryonic development despite the ablation of DC. Interestingly, the DC-ablated dams had more foetal vessels, which could be a compensatory reaction to the reduced blood flow due to the impaired spiral artery modification to maintain the foetal supply. Furthermore, the branching of maternal and foetal vessels and the placental depth are important characteristics for a normal placental development and are associated with pregnancy complications like preeclampsia [170,171]. In mice with ablated DC, supplementation with gal-1 restored the blood vessel branches and placental size. In conclusion, we showed that gal-1 contributes to early angiogenic processes and rescues the gestation in DC-ablated mice.

The newly described pro-angiogenic function of gal-1 during gestation prompted us to investigate the involvement of this lectin in the outcome of pregnancy. We used two approaches to define the pregnancy progression in the absence of gal-1 *in vivo*. First, we injected the synthetic peptide anginex to block the pro-angiogenic function of gal-1 during the implantation and placentation period. Gal-1 was previously described as a receptor for anginex [97] and we confirmed that *Lgals1* deficient dams did not respond to anginex treatment. Second, we used *Lgals1* deficient mice, which are fertile and display no reduction in the number of implantations when compared to their *Lgals1* wt counterparts [52]. Although the pregnancy progression was not precluded in anginex-treated or *Lgals1* deficient dams, these mice developed PE-like symptoms including hypertension and renal dysfunction with proteinuria. In this context, the release of anti-angiogenic factors from the placenta resulted in dysfunctional endothelia, as we observed in the kidney, and a systemic inflammatory response, which both contribute to the manifestation of hypertension [114]. sFlt-1 and sEng are two of these bioactive factors and are elevated in PE patients as well as in rat and mouse models of PE [119,120,172]. We detected elevated sEng levels in the circulation of anginex-treated dams when compared to control mice. In the *Lgals1* deficient females, sFlt-1 and sEng were increased compared to their *Lgals1* wt counterparts, but did not reach significance, which might be explained by the different genetic background of the animals (129/P3J vs. C57BL/6J) in the anginex experiment. Nevertheless, sFlt-1 binds to PGF and VEGF thus preventing their interaction with endothelial cells [173], and the reduction in placental PGF and VEGF might be a result of elevated sFlt-1 in *Lgals1* deficient dams.

The activation of circulating and local RAS also contributes to endothelial dysfunctions [168]. Thereby RAS components are dysregulated and favour the development of hypertension during the course of PE. We observed an increase in placental renin and AT1 as also reported for human PE patients [174,175]. Furthermore, AT1AA are known to activate the RAS and thus contribute to the pathogenesis of PE [144]. Interestingly, AT1AA induces PE-like symptoms in mice [176] and we detected increased levels of AT1AA in the circulation of anginex-treated and *Lgals1* deficient dams. These results were supported by our microarray analysis of human extravillous trophoblasts cells. The inhibition of gal-1 by anginex treatment *in vitro* resulted in the dysregulation of genes that are involved in the development of hypertension,

Discussion

endothelial functions and RAS regulation (e.g. BMPER, ENG, ANGPTL2, AT1). We thus showed that gal-1 influences processes that contribute to PE pathogenesis.

During PE, the main threat for the unborn child is a compromised placental development resulting in IUGR [177]. We showed that gal-1 inhibition or deficiency resulted in foetuses with reduced body weights and a delayed development suggesting that they suffered from IUGR. In line with this, we observed smaller placentas with fewer glycogen cells, indicating a decreased availability of nutrients [17]. These mice also showed a reduction in tissue-associated uNK cells in the maternal decidua, which might account for a compromised spiral artery remodelling and thus decreased foetal blood supply. In humans, severe PE is associated with defects in the differentiation of cytotrophoblasts [178]. Consequently, PE placentas display a reduced interstitial and endovascular invasion by extravillous CTB thus preventing the adequate remodelling of maternal spiral arteries into low resistant, dilated blood vessels. In concordance with this, we observed thicker arterial walls upon gal-1 inhibition *in vivo*.

Additionally, our *in vitro* studies with human extravillous trophoblasts showed that gal-1 inhibition compromised the tube formation and branching capability in these cells. Most importantly, the adhesion to endothelial cells, which is a main step in the remodelling of spiral arteries by extravillous CTB, was diminished. In line with this, microarray studies of these cells revealed the dysregulation of genes that are associated with trophoblast invasion, differentiation, proliferation and tube formation (e.g. C1QTNFs, IGFBP1, FST, PSGs, OXTR, ANGPTL4).

A recent study showed that the dysregulation of genes in CTB of PE placentas is normalised after 48 h cultivation; it appears that the gene expression is dependent on the *in vivo* environment [179]. Since extravillous CTB express NRP-1 [87] and further *in vitro* studies showed that gal-1 regulates the invasion and differentiation of trophoblasts [41,48,180], we argue that disruption of gal-1–NRP-1 signalling results in a shallow utero-placental invasion and poor maternal artery remodelling leading to a placental hypoperfusion. As a consequence, PE placentas are characterised by hypoxia and oxidative stress [114,116], which we detected in the absence of gal-1 *in vivo* and *in vitro*. In this context, the protein expression of placental lactogen-I was down-regulated under hypoxic conditions *in vitro* [181] and proliferin-expressing trophoblasts are tightly associated with spiral artery modifications in mice [12]. We observed indeed a reduction of these placental hormones upon gal-1 inhibition *in vivo*. Therefore, we hypothesise that the inhibition or absence of gal-1 accounts for the compromised placental development and the development of PE-like symptoms in the anginex-treated and *Lgals1* deficient dams.

Pregnancy complications have previously been associated with a dysregulated expression of circulating and placental gal-1 in humans [41,104]. Additionally, women suffering from PE display fewer gal-1-expressing NK and T cells in their periphery, which may cause the activation of immune cells [133]. This is supported by the fact that PE is characterised by an excessive inflammatory response [182]. In line with this, the microarray analysis of human extravillous trophoblasts demonstrated that inflammatory mediators such as interleukins and their receptors are up-regulated upon gal-1 inhibition by anginex. Microarray analysis of the utero-placental tissue of preeclamptic women further emphasised the unique role of gal-1 in the pathogenesis of PE, as it was the only down-regulated galectin of this protein family.

When analysing the contribution of gal-1 to the PE pathogenesis, the discrimination between early and late onset PE is essential. Early PE placentas displayed a reduced gal-1 expression, while the placentas of women with late PE showed higher levels of gal-1. Consistently, higher gal-1 levels have previously been reported in late PE placentas [105]. We thus underline the importance of classifying PE patients according to the severity of the disease and support the hypothesis that early and late onset PE have different aetiologies [114]. In late PE patients, the maternal adaptions to gestation are disturbed. Due to the later onset (> 34 weeks of gestation) of the disease and the rather mild symptoms, the clinical management is facilitated both for the mother and the unborn child. We provide evidence that gal-1 is a pro-angiogenic factor during early and mid-gestation. We further argue that the increased gal-1 expression in the placental villi of late PE patients is a response to the reduced placental perfusion. Gal-1 thus promotes angiogenic processes in order to maintain an adequate blood supply of the foetus. Moreover, gal-1 is known to regulate tolerogenic processes during pregnancy [52,63,65] and may be present to counteract the excessive inflammatory response seen during PE [182].

In contrast, the clinical management in early PE patients is hindered, because of the early onset of the disease (< 34 weeks of gestation) and the severe symptoms. The hypertension and endothelial dysfunctions can be life-threatening for the mother and often requires a preterm delivery of the baby, whose development, especially of the lung, is premature and which already suffers from IUGR. We suggest that the reduced gal-1 levels might be one cause of the markedly compromised placental development in early PE women. Considering our results that gal-1 has pro-angiogenic functions during placentation *in vivo* and is involved in the migration and adhesion of human extravillous CTB *in vitro*, a reduction in gal-1 may have detrimental effects on placental development. This includes an impaired spiral artery remodelling and thus a reduced utero-placental blood-flow, which is associated with the manifestation of the early PE syndrome [116].

Finally, the circulating gal-1 levels were dysregulated in women suffering from PE and further emphasised that the discrimination between early and late PE is necessary when diagnosing this syndrome. For instance, we showed that gal-1 was increased only in late PE patients in the third trimester of pregnancy. Most interesting regarding the potential clinical use of our results is the reduction of gal-1 during the second trimester of late PE

Discussion

women, when they displayed no clinical signs of PE. In this context, the VEGFR2 ectodomain shedding, which can regulate the receptor function and is induced by specific ligands, might play a role. In PE patients, reduced VEGFR2 levels were observed in the plasma when compared to healthy control women [183]. First, lower VEGFR2 levels are a result of a reduced shedding from endothelial cells due to endothelial dysfunctions, which are characteristic for PE. Second, the reduced bioavailability of VEGF, caused by increased circulating sFlt-1 during PE, is known to decrease VEGFR2 synthesis and trafficking on endothelial cells and thus also contributes to the reduction of plasma VEGFR2. Knowing that gal-1 binding to NRP-1 enhances VEGFR2 signalling [88], we hypothesise that the lowered gal-1 levels in PE patients also decrease the ligand-induced shedding of VEGFR2 and thus reduce plasma VEGFR2. The reduction in circulating gal-1 during the second trimester may thus contribute to a reduced VEGFR2 signalling and subsequent PE pathogenesis. We identified gal-1 as a valuable biomarker for early PE diagnosis in second trimester screenings. This offers the possibility for closer surveillance of high-risk patients before the onset of the first clinical symptoms.

With the present thesis, we aimed to elucidate the role of gal-1 in the angiogenic processes during gestation and its possible involvement in the pathogenesis of preeclampsia. We recently published the main results showing that gal-1 is required for healthy gestation and acts by supporting early angiogenic processes and placental development. Our paper also describes that gal-1 function could contribute to the PE pathogenesis [184]. For the first time, we described the pro-angiogenic function of gal-1 in pregnancy using a mouse model with a reduced vascular expansion during early gestation. We showed that gal-1 was able to overcome the implantation failure in mice with ablated DC and to rescue gestation by supporting a normal placentation and foetal development until term. Furthermore, interference with gal-1 functions resulted in compromised trophoblast properties *in vitro* and led to PE-like symptoms *in vivo*. Anginex-treated and *Lgals1* deficient mice suffered from hypertension, proteinuria, endothelial dysfunctions, dysregulated RAS, placental hypoperfusion, and IUGR implying an important role for gal-1 in the pathogenesis of PE. Finally, we showed that gal-1 is differentially expressed in the placenta of healthy pregnant and early and late onset PE patients in the third trimester of pregnancy, highlighting the importance of distinguishing between early (severe) and late (mild) PE. Additionally, we showed that circulating levels of gal-1 could be employed as valuable biomarker for the early diagnosis of PE. However, there is the need of a future prospective study in which patients are classified regarding early and late onset of the disease. The circulating gal-1 levels should be monitored before the onset of any symptoms, so that the definition of gal-1 as a PE biomarker can be completed.

5. Summary

Preeclampsia (PE) is a multisystemic disorder affecting about 12 to 22% of pregnant women and is the main cause for maternal and foetal morbidity and mortality. The clinical onset of the disease occurs in the second half of pregnancy (> 20 weeks) and the only ultimate cure is the delivery of the baby. The typical symptoms consist of hypertension and proteinuria, mainly caused by endothelial dysfunctions and the release of anti-angiogenic factors from the placenta into the maternal circulation. The pathogenesis of PE is associated with an excessive maternal inflammatory response and poor placentation. Due to an impaired remodelling of the maternal spiral arteries, the utero-placental blood flow is compromised and thus causes intrauterine growth restriction of the foetus.

Galectin-1 (gal-1) is a prototypic member of the evolutionary conserved family of galactoside-binding lectins. Gal-1 is abundant in the female reproductive tract and its expression increases during mammalian pregnancy. It was previously shown that gal-1 regulates maternal tolerance towards foetal antigens during pregnancy and promotes angiogenic processes during pathological conditions like cancer. Here, we demonstrated that gal-1 promotes decidual vascular expansion through neuropilin-1 (NRP-1) / vascular endothelial growth factor receptor 2 (VEGFR2) signalling and thus exhibits pro-angiogenic functions during early pregnancy. Gal-1 was able to rescue the placentation process in mice with ablated dendritic cells. Also in mice, the inhibition of or deficiency in gal-1 caused a spontaneous PE-like syndrome leading to hypertension, proteinuria and endothelial dysfunctions. Gal-1 inhibition impaired trophoblast functions that are necessary for a good placentation and spiral artery remodelling *in vivo* and *in vitro*. In line with this, we detected dysregulated circulating and placental levels of gal-1 in women that suffer from PE. Thus, we showed that the pro-angiogenic function of gal-1 is required for a healthy gestation and propose gal-1 as a potential biomarker for the early diagnosis of PE in humans.

Zusammenfassung

Präeklampsie (PE) ist eine multisystemische Erkrankung und entwickelt sich bei etwa 12 bis 22% aller schwangeren Frauen. Sie ist der Hauptgrund für maternale und fötale Morbidität und Mortalität. Der Krankheitsausbruch erfolgt in der zweiten Schwangerschaftshälfte (> 20 Wochen) und die einzige Heilung ist die Entbindung des Kindes. Die typischen Symptome umfassen Bluthochdruck (Hypertonie) und die Ausscheidung von Eiweißen im Urin (Proteinurie), die hauptsächlich durch endotheliale Fehlfunktionen und die Freisetzung von anti-angiogenen Faktoren aus der Plazenta in die maternale Zirkulation ausgelöst werden. Die Pathogenese von PE steht mit einer überschießenden imflammatorischen Immunantwort der Mutter und einer schlechten Plazentation in Zusammenhang. Durch eine verminderte Remodellierung der maternalen Spiralarterien ist der utero-plazentale Blutfluss eingeschränkt und bewirkt somit eine intrauterine Wachstumsretardierung des Fötus.

Galektin-1 (Gal-1) ist ein typisches Mitglied der evolutionär konservierten Familie der Galaktosid-bindenden Lektine. Gal-1 kommt in großer Menge in den weiblichen Reproduktionsorganen vor und seine Expression steigt während der Schwangerschaft in Säugetieren weiter an. Es wurde bereits gezeigt, dass Gal-1 die maternale Toleranz gegenüber fötalen Antigenen reguliert und angiogene Prozesse unter pathologischen Bedingungen wie zum Beispiel bei Krebserkrankungen fördert. In dieser Arbeit zeigen wir, dass Gal-1 die vaskuläre Expansion in der Dezidua fördert und über den Neuropilin-1 (NRP-1) / vaskulärer endothelialer Wachstumsfaktor-Rezeptor 2 (VEGFR2) – Signalweg vermittelt. Gal-1 besitzt somit eine pro-angiogene Wirkung während der frühen Schwangerschaft. In Mäusen mit verminderten dendritischen Zellen konnte Gal-1 den normalen Plazentationsprozess wiederherstellen. Die Gal-1 – Inhibierung oder – Defizienz führte in Mäusen zu einem spontanen PE-ähnlichem Syndrom mit Hypertonie, Proteinurie und endothelialen Fehlfunktionen. Die Inhibierung von Gal-1 beeinträchtigt die

Summary

Funktion von Trophoblasten, die für eine gute Plazentation und Remodellierung der Spiralarterien *in vivo* und *in vitro* notwendig sind. Im Einklang mit diesen Ergebnissen entdeckten wir deregulierte Gal-1 – Level in der Zirkulation und Plazenta von Frauen, die an PE leiden. Wir haben damit gezeigt, dass die pro-angiogene Funktion von Gal-1 für eine gesunde Schwangerschaft notwendig ist und schlagen Gal-1 als einen potentiellen Biomarker für die frühe Diagnose von PE im Menschen vor.

6. References

1. Allen, E. The oestrous cycle in the mouse. *American Journal of Anatomy* **30**, 297-371 (1922).
2. Cross, J. C., Werb, Z. & Fisher, S. J. Implantation and the placenta: key pieces of the development puzzle. *Science* **266**, 1508-1518 (1994).
3. Paria, B. C., Huet-Hudson, Y. M. & Dey, S. K. Blastocyst's state of activity determines the "window" of implantation in the receptive mouse uterus. *Proceedings of the National Academy of Sciences of the United States of America* **90**, 10159-10162 (1993).
4. Glasser, S., Mulholland, J., Mani, S. & Julian, J. Blastocyst-endometrial relationships: reciprocal interactions between uterine epithelial and stromal cells and blastocysts. *Trophoblast Research* **5**, 229-229 (1991).
5. Ramathal, C. Y., Bagchi, I. C., Taylor, R. N. & Bagchi, M. K. Endometrial decidualization: of mice and men. *Seminars in reproductive medicine* **28**, 17-26, doi:10.1055/s-0029-1242989 (2010).
6. Rose, K. D. & Archibald, J. D. *The rise of placental mammals : origins and relationships of the Major Extant Clades.* (Johns Hopkins University Press, 2005).
7. Benirschke, K. & Kaufmann, P. (New York: Springer.-974p, 2000).
8. Malassine, A., Frendo, J. L. & Evain-Brion, D. A comparison of placental development and endocrine functions between the human and mouse model. *Human reproduction update* **9**, 531-539 (2003).
9. Moffett, A., Loke, Y. W. & McLaren, A. *Biology and pathology of trophoblast.* (Cambridge University Press, 2006).
10. Cross, J. C. *et al.* Trophoblast functions, angiogenesis and remodeling of the maternal vasculature in the placenta. *Molecular and cellular endocrinology* **187**, 207-212 (2002).
11. Ma, G. T. *et al.* GATA-2 and GATA-3 regulate trophoblast-specific gene expression in vivo. *Development* **124**, 907-914 (1997).
12. Adamson, S. L. *et al.* Interactions between trophoblast cells and the maternal and fetal circulation in the mouse placenta. *Developmental biology* **250**, 358-373 (2002).
13. He, Y. *et al.* Alternative splicing of vascular endothelial growth factor (VEGF)-R1 (FLT-1) pre-mRNA is important for the regulation of VEGF activity. *Molecular endocrinology* **13**, 537-545 (1999).
14. Jackson, D., Volpert, O. V., Bouck, N. & Linzer, D. I. Stimulation and inhibition of angiogenesis by placental proliferin and proliferin-related protein. *Science* **266**, 1581-1584 (1994).
15. Linzer, D. I. & Nathans, D. A new member of the prolactin-growth hormone gene family expressed in mouse placenta. *The EMBO journal* **4**, 1419-1423 (1985).
16. Ain, R., Canham, L. N. & Soares, M. J. Gestation stage-dependent intrauterine trophoblast cell invasion in the rat and mouse: novel endocrine phenotype and regulation. *Developmental biology* **260**, 176-190 (2003).
17. Coan, P. M., Conroy, N., Burton, G. J. & Ferguson-Smith, A. C. Origin and characteristics of glycogen cells in the developing murine placenta. *Developmental dynamics : an official publication of the American Association of Anatomists* **235**, 3280-3294, doi:10.1002/dvdy.20981 (2006).
18. Maruyama, T. & Yoshimura, Y. Molecular and cellular mechanisms for differentiation and regeneration of the uterine endometrium. *Endocrine journal* **55**, 795-810 (2008).
19. Wilcox, A. J., Baird, D. D. & Weinberg, C. R. Time of implantation of the conceptus and loss of pregnancy. *The New England journal of medicine* **340**, 1796-1799, doi:10.1056/NEJM199906103402304 (1999).
20. Aplin, J. D. Implantation, trophoblast differentiation and haemochorial placentation: mechanistic evidence in vivo and in vitro. *J Cell Sci* **99 (Pt 4)**, 681-692 (1991).
21. Schlafke, S. & Enders, A. C. Cellular basis of interaction between trophoblast and uterus at implantation. *Biol Reprod* **12**, 41-65 (1975).
22. James, J. L., Stone, P. R. & Chamley, L. W. Cytotrophoblast differentiation in the first trimester of pregnancy: evidence for separate progenitors of extravillous trophoblasts and syncytiotrophoblast. *Reproduction* **130**, 95-103, doi:10.1530/rep.1.00723 (2005).
23. Barondes, S. H. *et al.* Galectins: a family of animal beta-galactoside-binding lectins. *Cell* **76**, 597-598 (1994).

References

24 Lopez-Lucendo, M. F. et al. Growth-regulatory human galectin-1: crystallographic characterisation of the structural changes induced by single-site mutations and their impact on the thermodynamics of ligand binding. *Journal of molecular biology* **343**, 957-970, doi:10.1016/j.jmb.2004.08.078 (2004).
25 Cho, M. & Cummings, R. D. Galectin-1, a beta-galactoside-binding lectin in Chinese hamster ovary cells. I. Physical and chemical characterization. *J Biol Chem* **270**, 5198-5206 (1995).
26 Camby, I., Le Mercier, M., Lefranc, F. & Kiss, R. Galectin-1: a small protein with major functions. *Glycobiology* **16**, 137R-157R, doi:10.1093/glycob/cwl025 (2006).
27 Liu, F. T., Patterson, R. J. & Wang, J. L. Intracellular functions of galectins. *Biochim Biophys Acta* **1572**, 263-273 (2002).
28 Ozeki, Y. et al. Tissue fibronectin is an endogenous ligand for galectin-1. *Glycobiology* **5**, 255-261 (1995).
29 Moiseeva, E. P., Javed, Q., Spring, E. L. & de Bono, D. P. Galectin 1 is involved in vascular smooth muscle cell proliferation. *Cardiovascular research* **45**, 493-502 (2000).
30 Moiseeva, E. P., Williams, B., Goodall, A. H. & Samani, N. J. Galectin-1 interacts with beta-1 subunit of integrin. *Biochemical and biophysical research communications* **310**, 1010-1016 (2003).
31 Cooper, D. N. & Barondes, S. H. Evidence for export of a muscle lectin from cytosol to extracellular matrix and for a novel secretory mechanism. *The Journal of cell biology* **110**, 1681-1691 (1990).
32 Hughes, R. C. Secretion of the galectin family of mammalian carbohydrate-binding proteins. *Biochim Biophys Acta* **1473**, 172-185 (1999).
33 Nickel, W. Unconventional secretory routes: direct protein export across the plasma membrane of mammalian cells. *Traffic* **6**, 607-614, doi:10.1111/j.1600-0854.2005.00302.x (2005).
34 Than, N. G. et al. Emergence of hormonal and redox regulation of galectin-1 in placental mammals: implication in maternal-fetal immune tolerance. *Proceedings of the National Academy of Sciences of the United States of America* **105**, 15819-15824, doi:10.1073/pnas.0807606105 (2008).
35 Scafoglio, C. et al. Comparative gene expression profiling reveals partially overlapping but distinct genomic actions of different antiestrogens in human breast cancer cells. *Journal of cellular biochemistry* **98**, 1163-1184, doi:10.1002/jcb.20820 (2006).
36 Orso, F. et al. Activator protein-2gamma (AP-2gamma) expression is specifically induced by oestrogens through binding of the oestrogen receptor to a canonical element within the 5'-untranslated region. *The Biochemical journal* **377**, 429-438, doi:10.1042/BJ20031133 (2004).
37 Choe, Y. S. et al. Expression of galectin-1 mRNA in the mouse uterus is under the control of ovarian steroids during blastocyst implantation. *Molecular reproduction and development* **48**, 261-266, doi:10.1002/(SICI)1098-2795(199710)48:2<261::AID-MRD14>3.0.CO;2-0 (1997).
38 Hirota, Y. et al. Galectin-1 markedly reduces the incidence of resorptions in mice missing immunophilin FKBP52. *Endocrinology* **153**, 2486-2493, doi:10.1210/en.2012-1035 (2012).
39 Phillips, B. et al. Differential expression of two beta-galactoside-binding lectins in the reproductive tracts of pregnant mice. *Biol Reprod* **55**, 548-558 (1996).
40 von Wolff, M., Wang, X., Gabius, H. J. & Strowitzki, T. Galectin fingerprinting in human endometrium and decidua during the menstrual cycle and in early gestation. *Molecular human reproduction* **11**, 189-194, doi:10.1093/molehr/gah144 (2005).
41 Tirado-Gonzalez, I. et al. Galectin-1 influences trophoblast immune evasion and emerges as a predictive factor for the outcome of pregnancy. *Molecular human reproduction* **19**, 43-53, doi:10.1093/molehr/gas043 (2013).
42 Bevan, B. H., Kilpatrick, D. C., Liston, W. A., Hirabayashi, J. & Kasai, K. Immunohistochemical localization of a beta-D-galactoside-binding lectin at the human maternofetal interface. *The Histochemical journal* **26**, 582-586 (1994).
43 Poirier, F., Timmons, P. M., Chan, C. T., Guenet, J. L. & Rigby, P. W. Expression of the L14 lectin during mouse embryogenesis suggests multiple roles during pre- and post-implantation development. *Development* **115**, 143-155 (1992).
44 Hirabayashi, J. & Kasai, K. Human placenta beta-galactoside-binding lectin. Purification and some properties. *Biochemical and biophysical research communications* **122**, 938-944 (1984).

References

45 Fischer, I. et al. Stimulation of syncytium formation in vitro in human trophoblast cells by galectin-1. *Placenta* **31**, 825-832, doi:10.1016/j.placenta.2010.06.016 (2010).
46 Ramhorst, R. E. et al. Galectin-1 confers immune privilege to human trophoblast: implications in recurrent fetal loss. *Glycobiology* **22**, 1374-1386, doi:10.1093/glycob/cws104 (2012).
47 Vicovac, L., Jankovic, M. & Cuperlovic, M. Galectin-1 and -3 in cells of the first trimester placental bed. *Human reproduction* **13**, 730-735 (1998).
48 Kolundzic, N. et al. Galectin-1 is part of human trophoblast invasion machinery--a functional study in vitro. *PLoS One* **6**, e28514, doi:10.1371/journal.pone.0028514 (2011).
49 Sargent, I. L. Maternal and fetal immune responses during pregnancy. *Experimental and clinical immunogenetics* **10**, 85-102 (1993).
50 Chaouat, G., Petitbarat, M., Dubanchet, S., Rahmati, M. & Ledee, N. Tolerance to the foetal allograft? *American journal of reproductive immunology* **63**, 624-636, doi:10.1111/j.1600-0897.2010.00832.x (2010).
51 Hunt, J. S., Petroff, M. G., McIntire, R. H. & Ober, C. HLA-G and immune tolerance in pregnancy. *FASEB journal : official publication of the Federation of American Societies for Experimental Biology* **19**, 681-693, doi:10.1096/fj.04-2078rev (2005).
52 Blois, S. M. et al. A pivotal role for galectin-1 in fetomaternal tolerance. *Nature medicine* **13**, 1450-1457, doi:10.1038/nm1680 (2007).
53 Blois, S. M. et al. Depletion of CD8+ cells abolishes the pregnancy protective effect of progesterone substitution with dydrogesterone in mice by altering the Th1/Th2 cytokine profile. *Journal of immunology* **172**, 5893-5899 (2004).
54 Lin, H., Mosmann, T. R., Guilbert, L., Tuntipopipat, S. & Wegmann, T. G. Synthesis of T helper 2-type cytokines at the maternal-fetal interface. *Journal of immunology* **151**, 4562-4573 (1993).
55 Piccinni, M. P. et al. Defective production of both leukemia inhibitory factor and type 2 T-helper cytokines by decidual T cells in unexplained recurrent abortions. *Nature medicine* **4**, 1020-1024, doi:10.1038/2006 (1998).
56 Aluvihare, V. R., Kallikourdis, M. & Betz, A. G. Regulatory T cells mediate maternal tolerance to the fetus. *Nature immunology* **5**, 266-271, doi:10.1038/ni1037 (2004).
57 Sasaki, Y. et al. Decidual and peripheral blood CD4+CD25+ regulatory T cells in early pregnancy subjects and spontaneous abortion cases. *Molecular human reproduction* **10**, 347-353, doi:10.1093/molehr/gah044 (2004).
58 Somerset, D. A., Zheng, Y., Kilby, M. D., Sansom, D. M. & Drayson, M. T. Normal human pregnancy is associated with an elevation in the immune suppressive CD25+ CD4+ regulatory T-cell subset. *Immunology* **112**, 38-43, doi:10.1111/j.1365-2567.2004.01869.x (2004).
59 Sakaguchi, S. Regulatory T cells: key controllers of immunologic self-tolerance. *Cell* **101**, 455-458 (2000).
60 Toscano, M. A. et al. Differential glycosylation of TH1, TH2 and TH-17 effector cells selectively regulates susceptibility to cell death. *Nature immunology* **8**, 825-834, doi:10.1038/ni1482 (2007).
61 Ilarregui, J. M. et al. Tolerogenic signals delivered by dendritic cells to T cells through a galectin-1-driven immunoregulatory circuit involving interleukin 27 and interleukin 10. *Nature immunology* **10**, 981-991, doi:10.1038/ni.1772 (2009).
62 Friebe, A. et al. Neutralization of LPS or blockage of TLR4 signaling prevents stress-triggered fetal loss in murine pregnancy. *Journal of molecular medicine* **89**, 689-699, doi:10.1007/s00109-011-0743-5 (2011).
63 Koopman, L. A. et al. Human decidual natural killer cells are a unique NK cell subset with immunomodulatory potential. *J Exp Med* **198**, 1201-1212, doi:10.1084/jem.20030305 (2003).
64 Karimi, K. & Arck, P. C. Natural Killer cells: keepers of pregnancy in the turnstile of the environment. *Brain, behavior, and immunity* **24**, 339-347, doi:10.1016/j.bbi.2009.09.015 (2010).
65 Kopcow, H. D. et al. T cell apoptosis at the maternal-fetal interface in early human pregnancy, involvement of galectin-1. *Proceedings of the National Academy of Sciences of the United States of America* **105**, 18472-18477, doi:10.1073/pnas.0809233105 (2008).
66 Gonzalez, I. T. et al. Uterine NK cells are critical in shaping DC immunogenic functions compatible with pregnancy progression. *PLoS One* **7**, e46755, doi:10.1371/journal.pone.0046755 (2012).

References

67. Lin, Y. et al. TSLP-induced placental DC activation and IL-10(+) NK cell expansion: comparative study based on BALB/c x C57BL/6 and NOD/SCID x C57BL/6 pregnant models. *Clinical immunology* **126**, 104-117, doi:10.1016/j.clim.2007.09.006 (2008).
68. Hsi, B. L., Hunt, J. S. & Atkinson, J. P. Differential expression of complement regulatory proteins on subpopulations of human trophoblast cells. *Journal of reproductive immunology* **19**, 209-223 (1991).
69. Petroff, M. G. et al. B7 family molecules are favorably positioned at the human maternal-fetal interface. *Biol Reprod* **68**, 1496-1504, doi:10.1095/biolreprod.102.010058 (2003).
70. Moreau, P. et al. Molecular and immunologic aspects of the nonclassical HLA class I antigen HLA-G: evidence for an important role in the maternal tolerance of the fetal allograft. *American journal of reproductive immunology* **40**, 136-144 (1998).
71. Reynolds, L. P., Killilea, S. D. & Redmer, D. A. Angiogenesis in the female reproductive system. *FASEB journal : official publication of the Federation of American Societies for Experimental Biology* **6**, 886-892 (1992).
72. Klauber, N., Rohan, R. M., Flynn, E. & D'Amato, R. J. Critical components of the female reproductive pathway are suppressed by the angiogenesis inhibitor AGM-1470. *Nature medicine* **3**, 443-446 (1997).
73. Blois, S. M. et al. Early expression of pregnancy-specific glycoprotein 22 (PSG22) by trophoblast cells modulates angiogenesis in mice. *Biol Reprod* **86**, 191, doi:10.1095/biolreprod.111.098251 (2012).
74. Achen, M. G., Gad, J. M., Stacker, S. A. & Wilks, A. F. Placenta growth factor and vascular endothelial growth factor are co-expressed during early embryonic development. *Growth factors* **15**, 69-80 (1997).
75. Yotsumoto, S. et al. Expression of adrenomedullin, a hypotensive peptide, in the trophoblast giant cells at the embryo implantation site in mouse. *Developmental biology* **203**, 264-275, doi:10.1006/dbio.1998.9073 (1998).
76. Li, M. et al. Fetal-derived adrenomedullin mediates the innate immune milieu of the placenta. *The Journal of clinical investigation* **123**, 2408-2420, doi:10.1172/JCI67039 (2013).
77. Shweiki, D., Itin, A., Neufeld, G., Gitay-Goren, H. & Keshet, E. Patterns of expression of vascular endothelial growth factor (VEGF) and VEGF receptors in mice suggest a role in hormonally regulated angiogenesis. *The Journal of clinical investigation* **91**, 2235-2243, doi:10.1172/JCI116450 (1993).
78. Hess, A. P. et al. Decidual stromal cell response to paracrine signals from the trophoblast: amplification of immune and angiogenic modulators. *Biol Reprod* **76**, 102-117, doi:10.1095/biolreprod.106.054791 (2007).
79. Croy, B. A. et al. Imaging of vascular development in early mouse decidua and its association with leukocytes and trophoblasts. *Biol Reprod* **87**, 125, doi:10.1095/biolreprod.112.102830 (2012).
80. Degaki, K. Y., Chen, Z., Yamada, A. T. & Croy, B. A. Delta-like ligand (DLL)1 expression in early mouse decidua and its localization to uterine natural killer cells. *PLoS One* **7**, e52037, doi:10.1371/journal.pone.0052037 (2012).
81. Ashkar, A. A., Di Santo, J. P. & Croy, B. A. Interferon gamma contributes to initiation of uterine vascular modification, decidual integrity, and uterine natural killer cell maturation during normal murine pregnancy. *J Exp Med* **192**, 259-270 (2000).
82. Plaks, V. et al. Uterine DCs are crucial for decidua formation during embryo implantation in mice. *The Journal of clinical investigation* **118**, 3954-3965, doi:10.1172/JCI36682 (2008).
83. Barrientos, G. et al. CXCR4(+) dendritic cells promote angiogenesis during embryo implantation in mice. *Angiogenesis* **16**, 417-427, doi:10.1007/s10456-012-9325-6 (2013).
84. Krey, G. et al. In vivo dendritic cell depletion reduces breeding efficiency, affecting implantation and early placental development in mice. *Journal of molecular medicine* **86**, 999-1011, doi:10.1007/s00109-008-0379-2 (2008).
85. Halder, J. B. et al. Differential expression of VEGF isoforms and VEGF(164)-specific receptor neuropilin-1 in the mouse uterus suggests a role for VEGF(164) in vascular permeability and angiogenesis during implantation. *Genesis* **26**, 213-224 (2000).
86. Douglas, N. C. et al. Vascular endothelial growth factor receptor 2 (VEGFR-2) functions to promote uterine decidual angiogenesis during early pregnancy in the mouse. *Endocrinology* **150**, 3845-3854, doi:10.1210/en.2008-1207 (2009).

References

87 Baston-Buest, D. M. et al. Expression of the vascular endothelial growth factor receptor neuropilin-1 at the human embryo-maternal interface. *European journal of obstetrics, gynecology, and reproductive biology* **154**, 151-156, doi:10.1016/j.ejogrb.2010.10.018 (2011).
88 Hsieh, S. H. et al. Galectin-1, a novel ligand of neuropilin-1, activates VEGFR-2 signaling and modulates the migration of vascular endothelial cells. *Oncogene* **27**, 3746-3753, doi:10.1038/sj.onc.1211029 (2008).
89 Soker, S., Takashima, S., Miao, H. Q., Neufeld, G. & Klagsbrun, M. Neuropilin-1 is expressed by endothelial and tumor cells as an isoform-specific receptor for vascular endothelial growth factor. *Cell* **92**, 735-745 (1998).
90 Thijssen, V. L., Poirier, F., Baum, L. G. & Griffioen, A. W. Galectins in the tumor endothelium; opportunities for combined cancer therapy. *Blood* **110**, 2819-2827 (2007).
91 Croci, D. O. et al. Disrupting galectin-1 interactions with N-glycans suppresses hypoxia-driven angiogenesis and tumorigenesis in Kaposi's sarcoma. *J Exp Med* **209**, 1985-2000, doi:10.1084/jem.20111665 (2012).
92 Thijssen, V. L. et al. Tumor cells secrete galectin-1 to enhance endothelial cell activity. *Cancer Res* **70**, 6216-6224 (2010).
93 Ito, K. et al. Thiodigalactoside inhibits murine cancers by concurrently blocking effects of galectin-1 on immune dysregulation, angiogenesis and protection against oxidative stress. *Angiogenesis* **14**, 293-307 (2011).
94 Laderach, D. J. et al. A Unique Galectin Signature in Human Prostate Cancer Progression Suggests Galectin-1 as a Key Target for Treatment of Advanced Disease. *Cancer Res* **73**, 86-96, doi:10.1158/0008-5472.CAN-12-1260 (2013).
95 Schulkens, I. A., Griffioen, A. W. & Thijssen, V. L. in *ACS Symposium Series: Galectins and Disease Implications for Targeted Therapeutics*   (ed Anatole Klyosov) 233-247 (ACS Publications, 2012).
96 Poirier, F. & Robertson, E. J. Normal development of mice carrying a null mutation in the gene encoding the L14 S-type lectin. *Development* **119**, 1229-1236 (1993).
97 Thijssen, V. L. et al. Galectin-1 is essential in tumor angiogenesis and is a target for antiangiogenesis therapy. *Proceedings of the National Academy of Sciences of the United States of America* **103**, 15975-15980, doi:10.1073/pnas.0603883103 (2006).
98 Banh, A. et al. Tumor galectin-1 mediates tumor growth and metastasis through regulation of T-cell apoptosis. *Cancer Res* **71**, 4423-4431, doi:10.1158/0008-5472.CAN-10-4157 (2011).
99 Sorme, P. et al. Design and synthesis of galectin inhibitors. *Methods in enzymology* **363**, 157-169, doi:10.1016/S0076-6879(03)01050-4 (2003).
100 Dings, R. P., Nesmelova, I., Griffioen, A. W. & Mayo, K. H. Discovery and development of anti-angiogenic peptides: A structural link. *Angiogenesis* **6**, 83-91, doi:10.1023/B:AGEN.0000011730.94233.06 (2003).
101 Griffioen, A. W. et al. Anginex, a designed peptide that inhibits angiogenesis. *The Biochemical journal* **354**, 233-242 (2001).
102 van der Schaft, D. W. et al. The designer anti-angiogenic peptide anginex targets tumor endothelial cells and inhibits tumor growth in animal models. *FASEB journal : official publication of the Federation of American Societies for Experimental Biology* **16**, 1991-1993, doi:10.1096/fj.02-0509fje (2002).
103 Dings, R. P. et al. Anti-tumor activity of the novel angiogenesis inhibitor anginex. *Cancer Lett* **194**, 55-66 (2003).
104 Liu, A. X. et al. Proteomic analysis on the alteration of protein expression in the placental villous tissue of early pregnancy loss. *Biol Reprod* **75**, 414-420 (2006).
105 Than, N. G. et al. Severe preeclampsia is characterized by increased placental expression of galectin-1. *The journal of maternal-fetal & neonatal medicine : the official journal of the European Association of Perinatal Medicine, the Federation of Asia and Oceania Perinatal Societies, the International Society of Perinatal Obstet* **21**, 429-442, doi:10.1080/14767050802041961 (2008).
106 Jauniaux, E. & Burton, G. J. Pathophysiology of histological changes in early pregnancy loss. *Placenta* **26**, 114-123, doi:10.1016/j.placenta.2004.05.011 (2005).
107 Nagaishi, M. et al. Chromosome abnormalities identified in 347 spontaneous abortions collected in Japan. *The journal of obstetrics and gynaecology research* **30**, 237-241, doi:10.1111/j.1447-0756.2004.00191.x (2004).

References

108 Barnea, E. R., Hustin, J. & Jauniaux, E. *The First twelve weeks of gestation*. (Springer-Verlag, 1992).
109 Rai, R. & Regan, L. Recurrent miscarriage. *Lancet* **368**, 601-611, doi:10.1016/S0140-6736(06)69204-0 (2006).
110 Barrientos, G. *et al*. Involvement of galectin-1 in reproduction: past, present and future. *Human reproduction update*, doi:10.1093/humupd/dmt040 (2013).
111 Walker, J. J. Pre-eclampsia. *Lancet* **356**, 1260-1265, doi:10.1016/S0140-6736(00)02800-2 (2000).
112 Lain, K. Y. & Roberts, J. M. Contemporary concepts of the pathogenesis and management of preeclampsia. *JAMA : the journal of the American Medical Association* **287**, 3183-3186 (2002).
113 Koonin, L. M., MacKay, A. P., Berg, C. J., Atrash, H. K. & Smith, J. C. Pregnancy-related mortality surveillance--United States, 1987-1990. *MMWR. CDC surveillance summaries : Morbidity and mortality weekly report. CDC surveillance summaries / Centers for Disease Control* **46**, 17-36 (1997).
114 Redman, C. W. & Sargent, I. L. Latest advances in understanding preeclampsia. *Science* **308**, 1592-1594, doi:10.1126/science.1111726 (2005).
115 Ness, R. B. & Roberts, J. M. Heterogeneous causes constituting the single syndrome of preeclampsia: a hypothesis and its implications. *American journal of obstetrics and gynecology* **175**, 1365-1370 (1996).
116 Redman, C. W. & Sargent, I. L. Immunology of pre-eclampsia. *American journal of reproductive immunology* **63**, 534-543, doi:10.1111/j.1600-0897.2010.00831.x (2010).
117 Practice, A. C. o. O. ACOG practice bulletin. Diagnosis and management of preeclampsia and eclampsia. Number 33, January 2002. American College of Obstetricians and Gynecologists. *International journal of gynaecology and obstetrics: the official organ of the International Federation of Gynaecology and Obstetrics* **77**, 67-75 (2002).
118 Redman, C. W. & Sargent, I. L. Placental stress and pre-eclampsia: a revised view. *Placenta* **30 Suppl A**, S38-42, doi:10.1016/j.placenta.2008.11.021 (2009).
119 Maynard, S. E. *et al*. Excess placental soluble fms-like tyrosine kinase 1 (sFlt1) may contribute to endothelial dysfunction, hypertension, and proteinuria in preeclampsia. *The Journal of clinical investigation* **111**, 649-658, doi:10.1172/JCI17189 (2003).
120 Venkatesha, S. *et al*. Soluble endoglin contributes to the pathogenesis of preeclampsia. *Nature medicine* **12**, 642-649, doi:10.1038/nm1429 (2006).
121 Perkins, A. V. *et al*. Corticotrophin-releasing hormone and corticotrophin-releasing hormone binding protein in normal and pre-eclamptic human pregnancies. *British journal of obstetrics and gynaecology* **102**, 118-122 (1995).
122 Muttukrishna, S., Knight, P. G., Groome, N. P., Redman, C. W. & Ledger, W. L. Activin A and inhibin A as possible endocrine markers for pre-eclampsia. *Lancet* **349**, 1285-1288 (1997).
123 Mise, H. *et al*. Augmented placental production of leptin in preeclampsia: possible involvement of placental hypoxia. *The Journal of clinical endocrinology and metabolism* **83**, 3225-3229 (1998).
124 Olsson, M. G. *et al*. Increased levels of cell-free hemoglobin, oxidation markers, and the antioxidative heme scavenger alpha(1)-microglobulin in preeclampsia. *Free radical biology & medicine* **48**, 284-291, doi:10.1016/j.freeradbiomed.2009.10.052 (2010).
125 Torry, D. S., Wang, H. S., Wang, T. H., Caudle, M. R. & Torry, R. J. Preeclampsia is associated with reduced serum levels of placenta growth factor. *American journal of obstetrics and gynecology* **179**, 1539-1544 (1998).
126 Sabapatha, A., Gercel-Taylor, C. & Taylor, D. D. Specific isolation of placenta-derived exosomes from the circulation of pregnant women and their immunoregulatory consequences. *American journal of reproductive immunology* **56**, 345-355, doi:10.1111/j.1600-0897.2006.00435.x (2006).
127 Germain, S. J., Sacks, G. P., Sooranna, S. R., Sargent, I. L. & Redman, C. W. Systemic inflammatory priming in normal pregnancy and preeclampsia: the role of circulating syncytiotrophoblast microparticles. *Journal of immunology* **178**, 5949-5956 (2007).
128 Smarason, A. K., Sargent, I. L., Starkey, P. M. & Redman, C. W. The effect of placental syncytiotrophoblast microvillous membranes from normal and pre-eclamptic women on the growth of endothelial cells in vitro. *British journal of obstetrics and gynaecology* **100**, 943-949 (1993).

References

129 Jeschke, U. et al. Expression of galectin-1, -3 (gal-1, gal-3) and the Thomsen-Friedenreich (TF) antigen in normal, IUGR, preeclamptic and HELLP placentas. *Placenta* **28**, 1165-1173, doi:10.1016/j.placenta.2007.06.006 (2007).
130 Tal, R. The role of hypoxia and hypoxia-inducible factor-1Alpha in preeclampsia pathogenesis. *Biol Reprod* **87**, 134, doi:10.1095/biolreprod.112.102723 (2012).
131 Zhao, X. Y. et al. Hypoxia inducible factor-1 mediates expression of galectin-1: the potential role in migration/invasion of colorectal cancer cells. *Carcinogenesis* **31**, 1367-1375, doi:10.1093/carcin/bgq116 (2010).
132 Zhao, X. Y., Zhao, K. W., Jiang, Y., Zhao, M. & Chen, G. Q. Synergistic induction of galectin-1 by C/EBP{alpha} and HIF-1{alpha} and its role in differentiation of acute myeloid leukemic cells. *J Biol Chem* (2011).
133 Molvarec, A. et al. Peripheral blood galectin-1-expressing T and natural killer cells in normal pregnancy and preeclampsia. *Clinical immunology* **139**, 48-56, doi:10.1016/j.clim.2010.12.018 (2011).
134 Dabelea, D. et al. Increasing prevalence of gestational diabetes mellitus (GDM) over time and by birth cohort: Kaiser Permanente of Colorado GDM Screening Program. *Diabetes care* **28**, 579-584 (2005).
135 Hadar, E. & Hod, M. Establishing consensus criteria for the diagnosis of diabetes in pregnancy following the HAPO study. *Annals of the New York Academy of Sciences* **1205**, 88-93, doi:10.1111/j.1749-6632.2010.05671.x (2010).
136 Malcolm, J. Through the looking glass: gestational diabetes as a predictor of maternal and offspring long-term health. *Diabetes/metabolism research and reviews* **28**, 307-311, doi:10.1002/dmrr.2275 (2012).
137 Expert Committee on the, D. & Classification of Diabetes, M. Report of the expert committee on the diagnosis and classification of diabetes mellitus. *Diabetes care* **26 Suppl 1**, S5-20 (2003).
138 Bissonnette, J. M., Black, J. A., Wickham, W. K. & Acott, K. M. Glucose uptake into plasma membrane vesicles from the maternal surface of human placenta. *The Journal of membrane biology* **58**, 75-80 (1981).
139 Jansson, T., Wennergren, M. & Illsley, N. P. Glucose transporter protein expression in human placenta throughout gestation and in intrauterine growth retardation. *The Journal of clinical endocrinology and metabolism* **77**, 1554-1562 (1993).
140 Lee, M. Y. & Han, H. J. Galectin-1 upregulates glucose transporter-1 expression level via protein kinase C, phosphoinositol-3 kinase, and mammalian target of rapamycin pathways in mouse embryonic stem cells. *The international journal of biochemistry & cell biology* **40**, 2421-2430, doi:10.1016/j.biocel.2008.04.004 (2008).
141 Jung, S. et al. In vivo depletion of CD11c+ dendritic cells abrogates priming of CD8+ T cells by exogenous cell-associated antigens. *Immunity* **17**, 211-220 (2002).
142 Theiler, K. *The house mouse : atlas of embryonic development.* (Springer-Verlag, 1989).
143 Staff, A. C., Ranheim, T., Khoury, J. & Henriksen, T. Increased contents of phospholipids, cholesterol, and lipid peroxides in decidua basalis in women with preeclampsia. *American journal of obstetrics and gynecology* **180**, 587-592 (1999).
144 Wallukat, G. et al. Patients with preeclampsia develop agonistic autoantibodies against the angiotensin AT1 receptor. *The Journal of clinical investigation* **103**, 945-952, doi:10.1172/JCI4106 (1999).
145 Ahmed, A., Singh, J., Khan, Y., Seshan, S. V. & Girardi, G. A new mouse model to explore therapies for preeclampsia. *PLoS One* **5**, e13663, doi:10.1371/journal.pone.0013663 (2010).
146 Lai, Z., Kalkunte, S. & Sharma, S. A critical role of interleukin-10 in modulating hypoxia-induced preeclampsia-like disease in mice. *Hypertension* **57**, 505-514, doi:10.1161/HYPERTENSIONAHA.110.163329 (2011).
147 McCormick, J., Whitley, G. S., Le Bouteiller, P. & Cartwright, J. E. Soluble HLA-G regulates motility and invasion of the trophoblast-derived cell line SGHPL-4. *Human reproduction* **24**, 1339-1345, doi:10.1093/humrep/dep026 (2009).
148 Brazma, A. et al. Minimum information about a microarray experiment (MIAME)-toward standards for microarray data. *Nature genetics* **29**, 365-371, doi:10.1038/ng1201-365 (2001).
149 Benjamini, Y., Drai, D., Elmer, G., Kafkafi, N. & Golani, I. Controlling the false discovery rate in behavior genetics research. *Behavioural brain research* **125**, 279-284 (2001).

References

150 Ardi, V. C. *et al*. Neutrophil MMP-9 proenzyme, unencumbered by TIMP-1, undergoes efficient activation in vivo and catalytically induces angiogenesis via a basic fibroblast growth factor (FGF-2)/FGFR-2 pathway. *J Biol Chem* **284**, 25854-25866, doi:10.1074/jbc.M109.033472 (2009).

151 Hawinkels, L. J. *et al*. VEGF release by MMP-9 mediated heparan sulphate cleavage induces colorectal cancer angiogenesis. *Eur J Cancer* **44**, 1904-1913, doi:10.1016/j.ejca.2008.06.031 (2008).

152 Bajou, K. *et al*. Plasminogen activator inhibitor-1 protects endothelial cells from FasL-mediated apoptosis. *Cancer cell* **14**, 324-334, doi:10.1016/j.ccr.2008.08.012 (2008).

153 Leali, D. *et al*. Long pentraxin 3/tumor necrosis factor-stimulated gene-6 interaction: a biological rheostat for fibroblast growth factor 2-mediated angiogenesis. *Arteriosclerosis, thrombosis, and vascular biology* **32**, 696-703, doi:10.1161/ATVBAHA.111.243998 (2012).

154 Voronov, E. *et al*. IL-1 is required for tumor invasiveness and angiogenesis. *Proceedings of the National Academy of Sciences of the United States of America* **100**, 2645-2650, doi:10.1073/pnas.0437939100 (2003).

155 Jin, X. *et al*. Matriptase activates stromelysin (MMP-3) and promotes tumor growth and angiogenesis. *Cancer science* **97**, 1327-1334, doi:10.1111/j.1349-7006.2006.00328.x (2006).

156 Holloway, D. E., Chavali, G. B., Hares, M. C., Subramanian, V. & Acharya, K. R. Structure of murine angiogenin: features of the substrate- and cell-binding regions and prospects for inhibitor-binding studies. *Acta crystallographica. Section D, Biological crystallography* **61**, 1568-1578, doi:10.1107/S0907444905029616 (2005).

157 Russell, D. L., Doyle, K. M., Ochsner, S. A., Sandy, J. D. & Richards, J. S. Processing and localization of ADAMTS-1 and proteolytic cleavage of versican during cumulus matrix expansion and ovulation. *J Biol Chem* **278**, 42330-42339, doi:10.1074/jbc.M300519200 (2003).

158 Lin, C. G. *et al*. CCN3 (NOV) is a novel angiogenic regulator of the CCN protein family. *J Biol Chem* **278**, 24200-24208, doi:10.1074/jbc.M302028200 (2003).

159 Ongusaha, P. P. *et al*. HB-EGF is a potent inducer of tumor growth and angiogenesis. *Cancer Res* **64**, 5283-5290, doi:10.1158/0008-5472.CAN-04-0925 (2004).

160 Isozaki, T. *et al*. Evidence for CXCL16 as a potent angiogenic mediator and endothelial progenitor cell chemotactic factor. *Arthritis and rheumatism*, doi:10.1002/art.37981 (2013).

161 Suganami, E. *et al*. Leptin stimulates ischemia-induced retinal neovascularization: possible role of vascular endothelial growth factor expressed in retinal endothelial cells. *Diabetes* **53**, 2443-2448 (2004).

162 Ohlsson, R. *et al*. PDGFB regulates the development of the labyrinthine layer of the mouse fetal placenta. *Developmental biology* **212**, 124-136, doi:10.1006/dbio.1999.9306 (1999).

163 Faria, T. N., Ogren, L., Talamantes, F., Linzer, D. I. & Soares, M. J. Localization of placental lactogen-I in trophoblast giant cells of the mouse placenta. *Biol Reprod* **44**, 327-331 (1991).

164 Dechend, R. *et al*. AT(1) receptor agonistic antibodies from preeclamptic patients cause vascular cells to express tissue factor. *Circulation* **101**, 2382-2387 (2000).

165 Dechend, R. *et al*. AT1 receptor agonistic antibodies from preeclamptic patients stimulate NADPH oxidase. *Circulation* **107**, 1632-1639, doi:10.1161/01.CIR.0000058200.90059.B1 (2003).

166 Siddiqui, A. H. *et al*. Angiotensin receptor agonistic autoantibody is highly prevalent in preeclampsia: correlation with disease severity. *Hypertension* **55**, 386-393, doi:10.1161/HYPERTENSIONAHA.109.140061 (2010).

167 Sezer, S. D. *et al*. VEGF, PIGF and HIF-1alpha in placentas of early- and late-onset pre-eclamptic patients. *Gynecological endocrinology : the official journal of the International Society of Gynecological Endocrinology* **29**, 797-800, doi:10.3109/09513590.2013.801437 (2013).

168 Herse, F. *et al*. Dysregulation of the circulating and tissue-based renin-angiotensin system in preeclampsia. *Hypertension* **49**, 604-611, doi:10.1161/01.HYP.0000257797.49289.71 (2007).

169 Roberts, J. M. & Lain, K. Y. Recent Insights into the pathogenesis of pre-eclampsia. *Placenta* **23**, 359-372, doi:10.1053/plac.2002.0819 (2002).

170 Dokras, A. *et al*. Severe feto-placental abnormalities precede the onset of hypertension and proteinuria in a mouse model of preeclampsia. *Biol Reprod* **75**, 899-907, doi:10.1095/biolreprod.106.053603 (2006).

References

171 Garcia-Gonzalez, M. A. *et al.* Pkd1 and Pkd2 are required for normal placental development. *PLoS One* **5**, doi:10.1371/journal.pone.0012821 (2010).

172 Saad, A. F. *et al.* Effects of Pravastatin on Angiogenic and Placental Hypoxic Imbalance in a Mouse Model of Preeclampsia. *Reproductive sciences*, doi:10.1177/1933719113492207 (2013).

173 Levine, R. J. *et al.* Circulating angiogenic factors and the risk of preeclampsia. *The New England journal of medicine* **350**, 672-683, doi:10.1056/NEJMoa031884 (2004).

174 Sowers, J. R. *et al.* Expression of renin and angiotensinogen genes in preeclamptic and normal human placental tissue. *Hypertension in pregnancy* **12**, 163-171 (1993).

175 Leung, P. S., Tsai, S. J., Wallukat, G., Leung, T. N. & Lau, T. K. The upregulation of angiotensin II receptor AT(1) in human preeclamptic placenta. *Molecular and cellular endocrinology* **184**, 95-102 (2001).

176 Zhou, C. C. *et al.* Angiotensin receptor agonistic autoantibodies induce pre-eclampsia in pregnant mice. *Nature medicine* **14**, 855-862, doi:10.1038/nm.1856 (2008).

177 Cox, P. & Marton, T. Pathological assessment of intrauterine growth restriction. *Best practice & research. Clinical obstetrics & gynaecology* **23**, 751-764, doi:10.1016/j.bpobgyn.2009.06.006 (2009).

178 McMaster, M. T., Zhou, Y. & Fisher, S. J. Abnormal placentation and the syndrome of preeclampsia. *Semin Nephrol* **24**, 540-547 (2004).

179 Zhou, Y. *et al.* Reversal of gene dysregulation in cultured cytotrophoblasts reveals possible causes of preeclampsia. *The Journal of clinical investigation* **123**, 2862-2872, doi:10.1172/JCI66966 (2013).

180 Fischer, I., Jeschke, U., Friese, K., Daher, S. & Betz, A. G. The role of galectin-1 in trophoblast differentiation and signal transduction. *Journal of reproductive immunology* **90**, 35-40, doi:10.1016/j.jri.2011.04.004 (2011).

181 Gultice, A. D., Selesniemi, K. L. & Brown, T. L. Hypoxia inhibits differentiation of lineage-specific Rcho-1 trophoblast giant cells. *Biol Reprod* **74**, 1041-1050, doi:10.1095/biolreprod.105.047845 (2006).

182 Redman, C. W., Sacks, G. P. & Sargent, I. L. Preeclampsia: an excessive maternal inflammatory response to pregnancy. *American journal of obstetrics and gynecology* **180**, 499-506 (1999).

183 Munaut, C. *et al.* Differential expression of Vegfr-2 and its soluble form in preeclampsia. *PLoS One* **7**, e33475, doi:10.1371/journal.pone.0033475 (2012).

184 Freitag, N. *et al.* Interfering with Gal-1-mediated angiogenesis contributes to the pathogenesis of preeclampsia. *Proceedings of the National Academy of Sciences of the United States of America* **110**, 11451-11456, doi:10.1073/pnas.1303707110 (2013).

7. Publication list

1. Barrientos G, **Freitag N**, Tirado-González I, Unverdorben L, Jeschke U, Thijssen VL, Blois SM. Involvement of galectin-1 in reproduction: past, present and future. *Human reproduction update*, doi:10.1093/humupd/dmt040 (2013).
2. Barrientos G*, Tirado-González I*, **Freitag N**, Kobelt P, Moschansky P, Klapp BF, Thijssen VL, Blois SM. CXCR4(+) dendritic cells promote angiogenesis during embryo implantation in mice. *Angiogenesis* **16**, 417-427, doi:10.1007/s10456-012-9325-6 (2013).
3. Blois SM, Piccioni F*, **Freitag N***, Tirado-González I, Moschansky P, Lloyd R, Hensel-Wiegel K, Rose M, Garcia MG, Alaniz LD, Mazzolini G. Dendritic cells regulate angiogenesis associated with liver fibrogenesis. *Angiogenesis* **17**, 119-128, doi:10.1007/s10456-013-9382-5 (2014).
4. Blois SM*, Sulkowski G*, Tirado-González I, Warren J, **Freitag N**, Klapp BF, Rifkin D, Fuss I, Strober W, Dveksler GS. Pregnancy-specific glycoprotein 1 (PSG1) activates TGF-beta and prevents dextran sodium sulfate (DSS)-induced colitis in mice. *Mucosal immunology*, doi:10.1038/mi.2013.53 (2013).
5. Blois SM*, Tirado-González I*, Wu J, Barrientos G, Johnson B, Warren J, **Freitag N**, Klapp BF, Irmak S, Ergun S, Dveksler GS. Early expression of pregnancy-specific glycoprotein 22 (PSG22) by trophoblast cells modulates angiogenesis in mice. *Biol Reprod* **86**, 191, doi:10.1095/biolreprod.111.098251 (2012).
6. **Freitag N**, Tirado-González I, Barrientos G, Herse F, Thijssen VL, Weedon-Fekjær SM, Schulz H, Wallukat G, Klapp BF, Nevers T, Sharma S, Staff AC, Dechend R, Blois SM. Interfering with Gal-1-mediated angiogenesis contributes to the pathogenesis of preeclampsia. *Proceedings of the National Academy of Sciences of the United States of America* **110**, 11451-11456, doi:10.1073/pnas.1303707110 (2013).
7. González IT*, Barrientos G*, **Freitag N**, Otto T, Thijssen VL, Moschansky P, von Kwiatkowski P, Klapp BF, Winterhager E, Bauersachs S, Blois SM. Uterine NK cells are critical in shaping DC immunogenic functions compatible with pregnancy progression. *PloS One* **7**, e46755, doi:10.1371/journal.pone.0046755 (2012).
8. Heusschen R*, **Freitag N***, Tirado-González I, Barrientos G, Moschansky P, Muñoz-Fernández R, Leno-Durán E, Klapp BF, Thijssen VL, Blois SM. Profiling Lgals9 splice variant expression at the fetal-maternal interface: implications in normal and pathological human pregnancy. *Biol Reprod* **88**, 22, doi:10.1095/biolreprod.112.105460 (2013).
9. Tirado-González I, **Freitag N**, Barrientos G, Shaikly V, Nagaeva O, Strand M, Kjellberg L, Klapp BF, Mincheva-Nilsson L, Cohen M, Blois SM. Galectin-1 influences trophoblast immune evasion and emerges as a predictive factor for the outcome of pregnancy. *Molecular human reproduction* **19**, 43-53, doi:10.1093/molehr/gas043 (2013).

* These authors have contributed equally to the work.

8. Appendix

8.1 Abbreviations

aDC	ablated dendritic cells
AngII	angiotensin II
AT1(AA)	angiotensin II receptor type 1 (autoantibodies)
BSA	serum bovine albumin
CMFDA	5-chloromethylfluorescein diacetate
CTB	cytotrophoblast(s)
(c)DNA	(copy) deoxyribonucleic acid
(c/m)RNA	(copy/messenger) ribonucleic acid
CXCR4	C-X-C chemokine receptor type 4
DAPI	4',6-diamidino-2-phenylindole
DBA	Dolichos biflorus agglutinin
DC	dendritic cell(s)
DEPC	diethyl pyrocarbonate
DT(R)	Diphtheria toxin (receptor)
DTT	dithiothreitol
ECL	enhanced chemiluminescence
EDTA	ethylenediaminetetraacetic acid
EtOH	ethanol
FCS	foetal calf serum
FITC	fluorescein isothiocyanate
gal-1	galectin-1
gd	gestation day
GC	trophoblast giant cell(s)
GDM	gestational diabetes mellitus
GLUT	glucose transporter
hCG	human chorionic gonadotropin
HCl	hydrochloride
HIF-2α	Hypoxia-inducible factor-2α
HRP	horseradish peroxidase
IB4	Isolectin B4
ICM	inner cell mass
IFN-γ	interferon-γ
IgG	immunoglobulin

Appendix

i.p.	intraperitoneal
IUGR	intrauterine growth restriction
i.v.	intravenous
MMPs	matrix metallopeptidases
NK cell	natural killer cell
NRP-1	neuropilin-1
OD	optical density
PAS	Periodic acid-Schiff
PBS	phosphate buffered saline
PE	preeclampsia
PECAM-1	platelet endothelial cell adhesion molecule-1; CD31
PGF	placental growth factor
PL-I	placental lactogen-I
PLF	proliferin
Prp	proliferin-related protein
RAS	renin-angiotensin system
RSA	recurrent spontaneous abortion
RT	room temperature
RT-PCR	reverse transcription polymerase chain reaction
SA	spontaneous abortion
sEng	soluble endoglin
sFlt-1	soluble fms-like tyrosine kinase-1, other name: sVEGFR1
STB	syncytiotrophoblast(s)
TBS	Tris-buffered saline
TGF-β1	transforming growth factor-β1
TMB	3,3',5,5'-Tetramethylbenzidine
TRITC	tetramethyl Rhodamine isothiocyanate
VEGF	vascular endothelial growth factor
VEGFR2	vascular endothelial growth factor receptor 2, other name: Flk-1
β-ME	β-mercaptoethanol

Appendix

8.2 List of figures

Figure 1: Differentiation of trophoblast cells from the inner cell mass (ICM) and trophectoderm in mice.. 4

Figure 2: Placentation in mice.. 4

Figure 3: Differentiation of human trophoblasts 7

Figure 4: The human placenta 7

Figure 5: Poor placentation during preeclampsia.. 18

Figure 6: Experimental design in a model of reduced vascular development and attenuated expansion and maturation (CD11c.DTR mouse model) 51

Figure 7: Gal-1 prevents implantation failure in mice with ablated DC .. 52

Figure 8: Ablation of DC during early pregnancy causes a reduced vascular development, which is prevented by gal-1 supplementation ... 53

Figure 9: Gal-1 boosts an angiogenic milieu in aDC mice 55

Figure 10: Gal-1 supports vascular development in the placenta of mice with ablated DC.. 56

Figure 11: Gal-1 facilitates the placental development affected by DC ablation.. 57

Figure 12: Placental lactogen-I and proliferin expression in giant cells upon gal-1 supplementation.. 58

Figure 13: Gal-1 favours pregnancy progression and embryonic development in aDC mice 59

Figure 14: Experimental mouse model to inhibit the pro-angiogenic function of gal-1 with anginex.. 60

Figure 15: Mice treated with anginex suffer from IUGR.. 61

Figure 16: Inhibiting gal-1 pro-angiogenic functions compromises placental development 62

Figure 17: Placental hormones are reduced by anginex treatment 62

Figure 18: Blocking gal-1 pro-angiogenic function influences the oxygen tension in placental tissues 63

Appendix

Figure 19: Gal-1 inhibition is correlated with higher circulating sEng levels and hypertension in late gestation 64

Figure 20: Renal dysfunction during late gestation is observed upon anginex treatment 65

Figure 21: Anginex-treated mice show dysregulated RAS pathways 66

Figure 22: Trophoblast functions are affected by anginex treatment in vitro 68

Figure 23: Anginex treatment provokes differential regulation of genes related to inflammation and angiogenesis in SGHPL-4 cells 69

Figure 24: Lgals1 deficient embryos suffer from IUGR 70

Figure 25: Placental hormone expression in Lgals1 deficient mice 71

Figure 26: Placental development is retarded in Lgals1 deficient dams 72

Figure 27: A PE-like syndrome characterised Lgals1 deficient dams.... 73

Figure 28: Systemic and placental RAS dysregulation is observed in Lgals1 deficient mice 74

Figure 29: Differential placental gal-1 expression in early and late PE.. 76

Figure 30: Circulating gal-1 levels predict the development of PE in human pilot studies 77

8.3 List of tables

Table 1: Trophoblast cell functions in the mouse placenta 4
Table 2: Treatment of CD11c.DTR female mice 32
Table 3: Clinical characteristics of the Oslo cohort 34
Table 4: Clinical characteristics of the Brown University cohort 35
Table 5: 15% polyacrylamide gel for the resolution of total isolated protein 39
Table 6: Mouse primer sequences for qPCR in placental tissue 42
Table 7: Human primer sequences for qPCR in placental and decidual tissue 44
Table 8: Gal-1 up-regulated proteins involved in the angiogenesis process during early pregnancy 54
Table 9: Angiogenic factors in the circulation of Lgals1 mice 72
Table 10: Galectin profile in healthy pregnant women and PE patients 75
Table 11: Theiler stage criteria for the analysis of embryonic development in mice vi
Table 12: Selected genes that were dysregulated in human trophoblast cells by gal-1 inhibition vii

8.4 Description of Theiler stages

Table 11: Theiler stage criteria for the analysis of embryonic development in mice
Theiler stage (TS) criteria were used to analyse the embryonic development of mice at gestation days (gd) 15.5 and 16.5. The vaginal plug was denoted as gd 0.5. The embryos were fixed in Bouin's solution and cleared in 70% ethanol before analysis.

Gd	TS	Embryo	Eye lids	Pinna	Fingers	Toes	Umbilical hernia	Skin
15.5	23		open	covers half of external auditory meatus	not parallel	not parallel	prominent	thin and smooth
	24		fusing	covers external auditory meatus	parallel	not parallel	disappearing, less prominent	thickened, formed wrinkles
16.5	25		fused	covers external auditory meatus	parallel	parallel	disappeared	thickened, formed wrinkles

8.5 Selection of dysregulated genes in human trophoblast microarray

Table 12: Selected genes that were dysregulated in human trophoblast cells by gal-1 inhibition
SGHPL-4 cells were untreated or treated with anginex for 5 or 24 h. The relative mRNA expression was analysed by microarray (fold change upper value: untreated vs. 5 h, lower value: untreated vs. 24 h anginex; positive: up-, negative: down-regulated gene).

Gene name and symbol	Gene bank accession number	Fold change	p-value	Reference
Preeclampsia and pregnancy-associated hypertension				
BMP binding endothelial regulator; BMPER	NM_133468.3	-1,33728 -2,13445	3,88E-05	1, 2
angiotensin II receptor, type 1, transcript variant 1; AGTR1	NM_000685.4	-1,52359 -2,10071	0,00019825	3
angiotensin II receptor, type 1, transcript variant 3; AGTR1	NM_004835.3	-1,63979 -2,00271	0,00179038	3
oxidized low density lipoprotein (lectin-like) receptor 1; OLR1	NM_002543.3	-1,17222 -1,63775	1,54E-05	4
endothelial PAS domain protein 1; EPAS1	NM_001430.3	1,4565 1,51043	0,00260902	5
chemokine (C-X-C motif) ligand 12 (stromal cell-derived factor 1), transcript variant 1; CXCL12	NM_199168.2	1,55663 1,57232	0,00028154	6
chemokine (C-X-C motif) ligand 12 (stromal cell-derived factor 1), transcript variant 2; CXCL12	NM_000609.4	1,00921 1,51461	0,00230238	6
endothelial cell growth factor 1 (platelet-derived); ECGF1	NM_001953.2	1,09628 1,53284	0,00027476	7
vascular endothelial growth factor A, transcript variant 2; VEGFA	NM_003376.4	1,67734 2,61923	0,00029818	8
vascular endothelial growth factor A, transcript variant 3; VEGFA	NM_001025367.1	1,6 1,94642	2,52E-06	8
vascular endothelial growth factor C ; VEGFC	NM_005429.2	1,03065 1,76212	9,04E-06	9
endoglin (Osler-Rendu-Weber syndrome 1); ENG	NM_000118.1	1,23034 2,12292	2,45E-06	10
hypoxia-inducible factor 1, alpha subunit (basic helix-loop-helix transcription factor),	NM_001530.2	1,47014 1,89607	0,00013840	11

Appendix

transcript variant 1; HIF1A				
hypoxia-inducible factor 1, alpha subunit (basic helix-loop-helix transcription factor), transcript variant 2; HIF1A	NM_181054.1	1,5424 2,0776	4,49E-05	11
pentraxin-related gene, rapidly induced by IL-1 beta; PTX3	NM_002852.2	-1,46398 -1,8736	7,01E-05	12-14
angiopoietin-like 2; ANGPTL2	NM_012098.2	2,03358 3,05047	1,32E-05	15, 16
placental growth factor; PGF	NM_002632.4	1,3842 1,96097	1,62E-06	8, 17, 18
pregnancy-associated plasma protein A, pappalysin 1; PAPPA	NM_002581.3	1,58659 2,4374	3,45E-06	17, 19
prostaglandin-endoperoxide synthase 1 (prostaglandin G/H synthase and cyclooxygenase), transcript variant 2; PTGS1	NM_080591.1	1,02791 1,54777	0,00015541	20
Placentation and trophoblast functions				
pregnancy specific beta-1-glycoprotein 3; PSG3	NM_021016.3	-1,00018 -1,96827	0,00127451	21
pregnancy specific beta-1-glycoprotein 5; PSG5	NM_002781.2	-1,09296 -1,95604	0,00080561	21
oxytocin receptor; OXTR	NM_000916.3	-2,07358 -2,74353	4,06E-07	22
insulin-like growth factor binding protein 1; IGFBP1	NM_001013029.1	1,24191 -1,32784	5,42E-05	23
follistatin, transcript variant FST317; FST	NM_006350.2	-1,83097 -2,40915	6,54E-08	24
follistatin, transcript variant FST344; FST	NM_013409.1	-1,90813 -2,32789	4,59E-05	24
pregnancy specific beta-1-glycoprotein 6, transcript variant 1; PSG6	NM_001031850.2	-1,00754 -1,65391	0,00242689	25
nerve growth factor (beta polypeptide); NGF	NM_002506.2	-1,3617 -1,63117	0,003157	26
brain-derived neurotrophic factor, transcript variant 4; BDNF	NM_001709.3	-1,60591 -1,32226	0,00294697	27
angiopoietin-like 4, transcript variant 1; ANGPTL4	NM_139314.1	2,07952 5,06186	1,14E-05	28
Vascular injury and inflammation				
insulin-like growth factor binding protein 7; IGFBP7	NM_001553.1	2,01 4,04E-05	4,04E-05	29
interleukin 7 receptor; IL7R	NM_002185.2	-1,6118 0,00038413	0,00038413	30-33
interleukin 32, transcript variant	NM_001012633.1	1,22846	0,001083	30-33

Appendix

		1,50421		
4; IL32				
interleukin 8; IL-8	NM_000584.2	2,27539 1,65293	0,00328961	30-33
interleukin 11; IL-11	NM_000641.2	1,81353 1,85378	0,00093608	30-33
interleukin 1, beta; IL1-B	NM_000576.2	2,10074 2,02671	0,00392865	30-33
interleukin 6 (interferon, beta 2); IL-6	NM_000600.1	2,13838 2,85802	0,00098341	30-33
interleukin-1 receptor-associated kinase 2; IRAK	NM_001570.3	2,36992 2,25116	2,64E-06	30-33
tumor necrosis factor receptor superfamily, member 11b; TNFRSF11B	NM_002546.3	-1,56782 -1,41602	4,26E-07	34, 35
tumor necrosis factor receptor superfamily, member 21 ; TNFRSF21	NM_014452.3	-1,04326 1,47811	0,00177314	34, 35
tumor necrosis factor receptor superfamily, member 25, transcript variant 10; TNFRSF25	NM_148973.1	1,142 1,55137	5,45E-06	34, 35
tumor necrosis factor receptor superfamily, member 10b, transcript variant 1; TNFRSF10B	NM_003842.3	1,21074 1,55663	1,98E-05	34, 35
Fas (TNF receptor superfamily, member 6), transcript variant 7; FAS	NM_152877.1	-1,09123 1,62536	4,16E-07	36
tumor necrosis factor, alpha-induced protein 8-like 3; TNFAIP8L3	NM_207381.2	1,10471 1,70653	3,08E-05	34, 35
C1q and tumor necrosis factor related protein 6, transcript variant 1; C1QTNF6	NM_031910.3	1,27553 1,70781	5,19E-07	37-41
C1q and tumor necrosis factor related protein 1; C1QTNF1	NM_198594.1	1,25953 2,23894	2,36E-05	37-41
tumor necrosis factor receptor superfamily, member 14 (herpesvirus entry mediator); TNFRSF14	NM_003820.2	-1,05545 1,81353	1,90E-05	34, 35
tumor necrosis factor, alpha-induced protein 6; TNFAIP6	NM_007115.2	2,29268 2,98552	1,18E-05	34, 35
transforming growth factor, beta receptor III; TGFBR3	NM_003243.2	1,06015 -1,77096	4,26E-07	34, 35
transforming growth factor, beta 2; TGFB2	NM_003238.1	-1,29715 -1,5711	8,85E-05	34, 35
TGFB-induced factor homeobox 1, transcript variant	NM_170695.2	1,63627 1,28747	0,0001552	34, 35

Appendix

1; TGIF1				
insulin-like growth factor binding protein 3, transcript variant 1; IGFBP3	NM_001013398.1	-1,07391 -1,738	4,11E-05	42, 43
insulin-like growth factor binding protein 3, transcript variant 2; IGFBP3	NM_000598.4	-1,26327 -1,70819	2,97E-05	42, 43

References

1. LaMarca, B. et al. Hypertension in response to autoantibodies to the angiotensin II type I receptor (AT1-AA) in pregnant rats: role of endothelin-1. *Hypertension* **54**, 905-909, doi:10.1161/HYPERTENSIONAHA.109.137935 (2009).
2. Helbing, T. et al. Kruppel-like factor 15 regulates BMPER in endothelial cells. *Cardiovascular research* **85**, 551-559, doi:10.1093/cvr/cvp314 (2010).
3. Herse, F. et al. Dysregulation of the circulating and tissue-based renin-angiotensin system in preeclampsia. *Hypertension* **49**, 604-611, doi:10.1161/01.HYP.0000257797.49289.71 (2007).
4. Chigusa, Y. et al. Decreased lectin-like oxidized LDL receptor 1 (LOX-1) and low Nrf2 activation in placenta are involved in preeclampsia. *The Journal of clinical endocrinology and metabolism* **97**, E1862-1870, doi:10.1210/jc.2012-1268 (2012).
5. Jarvenpaa, J. et al. Altered expression of angiogenesis-related placental genes in pre-eclampsia associated with intrauterine growth restriction. *Gynecological endocrinology : the official journal of the International Society of Gynecological Endocrinology* **23**, 351-355, doi:10.1080/09513590701350291 (2007).
6. Hwang, H. S., Kwon, H. S., Sohn, I. S., Park, Y. W. & Kim, Y. H. Increased CXCL12 expression in the placentae of women with pre-eclampsia. *European journal of obstetrics, gynecology, and reproductive biology* **160**, 137-141, doi:10.1016/j.ejogrb.2011.10.007 (2012).
7. Salama, R. H., Fathalla, M. M., Mekki, A. R. & Elsadek Bel, K. Implication of umbilical cord in preeclampsia. *Medical principles and practice : international journal of the Kuwait University, Health Science Centre* **20**, 124-128, doi:10.1159/000321212 (2011).
8. Andraweera, P. H., Dekker, G. A., Laurence, J. A. & Roberts, C. T. Placental expression of VEGF family mRNA in adverse pregnancy outcomes. *Placenta* **33**, 467-472, doi:10.1016/j.placenta.2012.02.013 (2012).
9. Srinivas, S. K., Morrison, A. C., Andrela, C. M. & Elovitz, M. A. Allelic variations in angiogenic pathway genes are associated with preeclampsia. *American journal of obstetrics and gynecology* **202**, 445 e441-411, doi:10.1016/j.ajog.2010.01.040 (2010).
10. Venkatesha, S. et al. Soluble endoglin contributes to the pathogenesis of preeclampsia. *Nature medicine* **12**, 642-649, doi:10.1038/nm1429 (2006).
11. Caniggia, I. et al. Hypoxia-inducible factor-1 mediates the biological effects of oxygen on human trophoblast differentiation through TGFbeta(3). *The Journal of clinical investigation* **105**, 577-587, doi:10.1172/JCI8316 (2000).
12. Cozzi, V. et al. PTX3 as a potential endothelial dysfunction biomarker for severity of preeclampsia and IUGR. *Placenta* **33**, 1039-1044, doi:10.1016/j.placenta.2012.09.009 (2012).
13. Zhou, P. et al. The expression of pentraxin 3 and tumor necrosis factor-alpha is increased in preeclamptic placental tissue and maternal serum. *Inflammation research : official journal of the European Histamine Research Society ... [et al.]* **61**, 1005-1012, doi:10.1007/s00011-012-0507-x (2012).
14. Hamad, R. R., Eriksson, M. J., Berg, E., Larsson, A. & Bremme, K. Impaired endothelial function and elevated levels of pentraxin 3 in early-onset preeclampsia. *Acta obstetricia et gynecologica Scandinavica* **91**, 50-56 (2012).

15. Guo, D. F. *et al.* Development of hypertension and kidney hypertrophy in transgenic mice overexpressing ARAP1 gene in the kidney. *Hypertension* **48**, 453-459, doi:10.1161/01.HYP.0000230664.32874.52 (2006).
16. Aoi, J. *et al.* Angiopoietin-like protein 2 is an important facilitator of inflammatory carcinogenesis and metastasis. *Cancer Res* **71**, 7502-7512, doi:10.1158/0008-5472.CAN-11-1758 (2011).
17. Pennings, J. L. A. *et al.* Integrative data mining to identify novel candidate serum biomarkers for pre-eclampsia screening. *Prenatal diagnosis* **31**, 1153-1159 (2011).
18. Weed, S. *et al.* Examining the correlation between placental and serum placenta growth factor in preeclampsia. *American journal of obstetrics and gynecology* **207**, 140 e141-146, doi:10.1016/j.ajog.2012.05.003 (2012).
19. Moslemi Zadeh, N., Naghshvar, F., Peyvandi, S., Gheshlaghi, P. & Ehetshami, S. PP13 and PAPP-A in the First and Second Trimesters: Predictive Factors for Preeclampsia? *ISRN obstetrics and gynecology* **2012**, 263871, doi:10.5402/2012/263871 (2012).
20. Wetzka, B. *et al.* Cyclooxygenase-1 and -2 in human placenta and placental bed after normal and pre-eclamptic pregnancies. *Human reproduction* **12**, 2313-2320 (1997).
21. Camolotto, S. *et al.* Expression and transcriptional regulation of individual pregnancy-specific glycoprotein genes in differentiating trophoblast cells. *Placenta* **31**, 312-319, doi:10.1016/j.placenta.2010.01.004 (2010).
22. Cassoni, P. *et al.* Activation of functional oxytocin receptors stimulates cell proliferation in human trophoblast and choriocarcinoma cell lines. *Endocrinology* **142**, 1130-1136 (2001).
23. Irwin, J. C., Suen, L. F., Martina, N. A., Mark, S. P. & Giudice, L. C. Role of the IGF system in trophoblast invasion and pre-eclampsia. *Human reproduction* **14 Suppl 2**, 90-96 (1999).
24. Bearfield, C., Jauniaux, E., Groome, N., Sargent, I. L. & Muttukrishna, S. The secretion and effect of inhibin A, activin A and follistatin on first-trimester trophoblasts in vitro. *European journal of endocrinology / European Federation of Endocrine Societies* **152**, 909-916, doi:10.1530/eje.1.01928 (2005).
25. Leslie, K. K. *et al.* Linkage of two human pregnancy-specific beta 1-glycoprotein genes: one is associated with hydatidiform mole. *Proceedings of the National Academy of Sciences of the United States of America* **87**, 5822-5826 (1990).
26. Kanai-Azuma, M. *et al.* Nerve growth factor promotes giant-cell transformation of mouse trophoblast cells in vitro. *Biochemical and biophysical research communications* **231**, 309-315, doi:10.1006/bbrc.1996.6032 (1997).
27. Kawamura, K. *et al.* Brain-derived neurotrophic factor promotes implantation and subsequent placental development by stimulating trophoblast cell growth and survival. *Endocrinology* **150**, 3774-3782, doi:10.1210/en.2009-0213 (2009).
28. Basak, S. & Duttaroy, A. K. cis-9,trans-11 conjugated linoleic acid stimulates expression of angiopoietin like-4 in the placental extravillous trophoblast cells. *Biochim Biophys Acta* **1831**, 834-843, doi:10.1016/j.bbalip.2013.01.012 (2013).
29. Liu, Z. K., Wang, R. C., Han, B. C., Yang, Y. & Peng, J. P. A novel role of IGFBP7 in mouse uterus: regulating uterine receptivity through Th1/Th2 lymphocyte balance and decidualization. *PLoS One* **7**, e45224, doi:10.1371/journal.pone.0045224 (2012).
30. Brewster, J. A. *et al.* Host inflammatory response profiling in preeclampsia using an in vitro whole blood stimulation model. *Hypertens Pregnancy* **27**, 1-16, doi:10.1080/10641950701826067 (2008).
31. Ramma, W. & Ahmed, A. Is inflammation the cause of pre-eclampsia? *Biochem Soc Trans* **39**, 1619-1627 (2011).
32. Borzychowski, A. M., Sargent, I. L. & Redman, C. W. G. Inflammation and pre-eclampsia. *Semin Fetal Neonatal Med* **11**, 309-316 (2006).
33. Redman, C. W. & Sargent, I. L. Latest advances in understanding preeclampsia. *Science* **308**, 1592-1594, doi:10.1126/science.1111726 (2005).
34. Cackovic, M. *et al.* Fractional excretion of tumor necrosis factor-alpha in women with severe preeclampsia. *Obstet Gynecol* **112**, 93-9100 (2008).

35 Keelan, J. A. & Mitchell, M. D. Placental cytokines and preeclampsia. *Front Biosci* **12**, 2706-2727 (2007).
36 Annells, M. F. *et al.* Interleukins-1, -4, -6, -10, tumor necrosis factor, transforming growth factor-beta, FAS, and mannose-binding protein C gene polymorphisms in Australian women: Risk of preterm birth. *American journal of obstetrics and gynecology* **191**, 2056-2067 (2004).
37 Agostinis, C. *et al.* MBL interferes with endovascular trophoblast invasion in pre-eclampsia. *Clin Dev Immunol* **2012**, 484321-484321 (2012).
38 Agostinis, C. *et al.* An alternative role of C1q in cell migration and tissue remodeling: contribution to trophoblast invasion and placental development. *Journal of immunology* **185**, 4420-4429 (2010).
39 Buurma, A. *et al.* Preeclampsia is characterized by placental complement dysregulation. *Hypertension* **60**, 1332-1337 (2012).
40 Girardi, G., Prohaszka, Z., Bulla, R., Tedesco, F. & Scherjon, S. Complement activation in animal and human pregnancies as a model for immunological recognition. *Mol Immunol* **48**, 1621-1630 (2011).
41 Singh, J., Ahmed, A. & Girardi, G. Role of complement component C1q in the onset of preeclampsia in mice. *Hypertension* **58**, 716-724 (2011).
42 Varma, M., de Groot, C. J., Lanyi, S. & Taylor, R. N. Evaluation of plasma insulin-like growth factor-binding protein-3 as a potential predictor of preeclampsia. *American journal of obstetrics and gynecology* **169**, 995-999 (1993).
43 Thadhani, R. *et al.* Insulin resistance and alterations in angiogenesis: additive insults that may lead to preeclampsia. *Hypertension* **43**, 988-992 (2004).

Acknowledgement

The research for this thesis was conducted in the Department of Psychosomatic Medicine, directed by Professor Burghard F. Klapp and later Dr Matthias Rose, at the Charité University Medicine in Berlin, Germany. I am grateful to the Fritz Thyssen Stiftung for supporting this project and the Commission for Young Scientific Investigators at the Charité for financing my dissertation through a pre-doctoral stipend.

Foremost, I would like to thank my supervisor Dr Sandra M. Blois for her dedication and encouragement during my doctoral studies. She helped me grow as a scientist and I deeply appreciate her support and feel honoured to be part of her group. I also want to thank Dr. Irene Tirado-González; it was a great pleasure to work with her. I would especially like to thank Petra Moschansky for her excellent technical assistance and kindness. I am very grateful to Dr Melanie L. Conrad for her valuable comments and suggestions on my thesis. In addition, I want to thank Dr Peter Kobelt, Petra Buße, Maria Daniltchenko, and Evelin Hagen for being great colleagues.

I would also like to thank Professor Fritz G. Rathjen who did not hesitate to become my second supervisor. A special thanks goes to our collaboration partners Dr Ralf Dechend and Dr Florian Herse (Experimental and Clinical Research Center, Berlin). Their ideas and helpful suggestions contributed greatly to my research project. I want to thank Dr Gerd Wallukat and Dr Herbert Schulz (Max Delbrück Center for Molecular Medicine, Berlin) for their experimental support. A special thanks goes to Professor Anne Cathrine Staff and Dr Susanne M. Weedon-Fekjær (Oslo University Hospital, Norway) as well as Professor Surendra Sharma and Tania Nevers (Brown University, Rhode Island, USA) for providing the human study materials. I also thank Professor

Victor L. J. L. Thijssen (VU University Medical Center, Amsterdam, Netherlands).

Finally, I would like to thank my parents who supported me in all my personal and professional decisions. Thanks for always being on my side. I want to thank my friend Joanna not only for informing me about the open Ph.D. position in the Blois lab, but also for being my dearest friend. I cannot express how grateful I am to my husband Elias.

I want morebooks!

Buy your books fast and straightforward online - at one of the world's fastest growing online book stores! Environmentally sound due to Print-on-Demand technologies.

Buy your books online at
www.get-morebooks.com

Kaufen Sie Ihre Bücher schnell und unkompliziert online – auf einer der am schnellsten wachsenden Buchhandelsplattformen weltweit! Dank Print-On-Demand umwelt- und ressourcenschonend produziert.

Bücher schneller online kaufen
www.morebooks.de

OmniScriptum Marketing DEU GmbH
Heinrich-Böcking-Str. 6-8
D - 66121 Saarbrücken
Telefax: +49 681 93 81 567-9

info@omniscriptum.com
www.omniscriptum.com

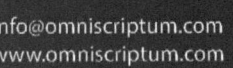

Printed by Books on Demand GmbH, Norderstedt / Germany